MANAGING SIX SIGMA

MANAGING SIX SIGMA

A Practical Guide to Understanding, Assessing, and Implementing the Strategy That Yields Bottom-Line Success

FORREST W. BREYFOGLE III
JAMES M. CUPELLO
BECKI MEADOWS
Smarter Solutions, Inc.
www.smartersolutions.com

A Wiley-Interscience Publication
JOHN WILEY & SONS, INC.
New York • Chichester • Weinheim • Brisbane • Singapore • Toronto

This book is printed on acid-free paper.

Published simultaneously in Canada.

This publication is designed to provide accurate and authoritative information in regard to the subject matter covered. It is sold with the understanding that the publisher is not engaged in rendering professional services. If professional advice or other expert assistance is required, the services of a competent professional person should be sought.

Library of Congress Cataloging-in-Publication Data:

Breyfogle, Forrest W., 1946-
    Managing Six Sigma : a practical guide to understanding, assessing, and implementing the strategy that yields bottom-line success / Forrest W. Breyfogle III, James M. Cupello, Becki Meadows.
        p.   cm.
        ISBN 0-471-39673-7 (cloth : alk. paper)
        1. Quality control—Statistical methods.   2. Production management—Statistical methods.
I. Cupello, James M. II. Meadows, Becki. III. Title.

TS156.B755 2000
658.5'62—62—dc21

                                                                                00-043301

Printed in the United States of America.

10 9 8 7 6 5 4 3 2

To my parents. Thank you for all your support. —F.B.

To my late father, Michael J. Cupello. His abundant knowledge never made it into print. He chose wife over wealth, fatherhood over fame, and responsibility over recognition. —J.C.

To my family, Prem, and especially Hans. Thank you for your patience, love, and support. —B.M.

CONTENTS

PREFACE xi

PART 1   WHY SIX SIGMA 1

1   How Six Sigma Compares to Other Quality Initiatives 3

1.1   What Is Six Sigma? 5
1.2   The Quality Leaders, 8
1.3   TQM and Six Sigma, 15
1.4   Six Sigma and Integration of the Statistical Tools, 17
1.5   Lean Manufacturing and Six Sigma, 21
1.6   Frequently Asked Questions about Six Sigma, 25

2   Six Sigma Background and Fundamentals 31

2.1   Motorola: The Birthplace of Six Sigma, 31
2.2   General Electric's Experiences with Six Sigma, 32
2.3   Eastman Kodak's Experience with Six Sigma, 33
2.4   Additional Experiences with Six Sigma, 35
2.5   Recent Advances, 36
2.6   The Statistical Definition of Six Sigma, 37

3   Six Sigma Needs Assessment 45

3.1   Implications of Quality Levels, 46
3.2   Cost of Poor Quality (COPQ), 47
3.3   Cost of Doing Nothing, 50
3.4   Assessment Questions, 50

PART 2 SIX SIGMA METRICS   53

4 Numbers and Information   55

    4.1   Example 4.1: Reacting to Data, 55

    4.2   Process Control Charting at the "30,000-Foot
Level," 59

    4.3   Discussion of Process Control Charting at the
"30,000-Foot Level," 61

    4.4   Control Charts at the "30,000-Foot Level": Attribute
Response, 65

    4.5   Goal Setting, Scorecard, and Measurements, 65

    4.6   Summary, 69

5 Crafting Insightful Metrics   71

    5.1   Six Sigma Metrics, 71

    5.2   Right Question, Right Metric, Right Activity, 78

    5.3   Example 5.1: Tracking Ongoing Product Compliance
from a Process Point of View, 81

    5.4   Example 5.2: Tracking and Improving Times for
Change Orders, 85

    5.5   Example 5.3: Improving the Effectiveness of
Employee Opinion Surveys, 87

    5.6   Example 5.4: Tracking and Reducing Overdue
Accounts Payable, 90

6 Performance Measurement   93

    6.1   Measurement Types, 93

    6.2   Principles of Measurement, 94

    6.3   The Balanced Scorecard, 97

PART 3 SIX SIGMA BUSINESS STRATEGY   99

7 Deployment Alternatives   101

    7.1   Deployment of Six Sigma: Prioritized Projects
with Bottom-Line Benefits, 102

    7.2   Advantages of Deploying Six Sigma through
Projects, 106

    7.3   Choosing a Six Sigma Provider, 107

    7.4   Essential Elements of an S^4 Implementation Plan , 109

8 Creating a Successful Six Sigma Infrastructure   115

    8.1   Executive Leadership, 117

    8.2   Customer Focus, 118

8.3   Strategic Goals, 121

8.4   Resources, 124

8.5   Metrics, 129

8.6   Culture, 130

8.7   Communications, 134

8.8   Lessons Learned, 135

9 Training and Implementation   139

9.1   Strategy, 140

9.2   Agendas for Six Sigma Training, 144

9.3   Computer Software, 146

9.4   Internal Versus External Training and Investment, 147

10 Project Selection, Sizing, and Other Techniques   149

10.1   Project Assessment, 150

10.2   Determine Metrics and Goals, 152

10.3   Determine a Baseline and Bottom-Line Benefits, 155

10.4   Scope the Project, 158

10.5   Create Project Charter and Overall Plan, 161

10.6   Motivated Team, 166

10.7   Project Execution Considerations, 171

10.8   Project Report-Outs to Upper Management, 174

10.9   Communicating and Leveraging Success, 179

PART 4 APPLYING SIX SIGMA   181

11 Manufacturing Applications   183

11.1   21-Step Integration of the Tools: Manufacturing Processes, 184

11.2   Example 11.1: Process Improvement and Exposing the Hidden Factory, 188

11.3   Example 11.2: Between- and Within-Part Variability, 191

11.4   Short Runs, 192

11.5   Engineering Process Control, 193

11.6   Example 11.3: Optimizing New Equipment Settings, 193

12 Service/Transactional Applications   195

12.1   Measuring and Improving Service/Transactional Processes, 196

12.2   21-Step Integration of the Tools: Service/Transactional
Processes, 197

12.3   Example 12.1: Improving On-Time Delivery, 200

12.4   Example 12.2: Applying DOE to Increase
Website Traffic, 205

12.5   Other Examples, 208

13   Development Applications   **211**

13.1   Measuring and Improving Development
Processes, 212

13.2   21-Step Integration of the Tools: Development
Processes, 213

13.3   Example 13.1: Notebook Computer
Development, 219

14   Need for Creativity, Invention, and Innovation   **221**

14.1   Definitions and Distinctions, 224

14.2   Encouraging Creativity, 230

List of Symbols   **239**

Glossary   **243**

References   **255**

INDEX   **263**

PREFACE

In recent years there has been much interest in the application of Six Sigma techniques. Within organizations chief executive officers (CEO) are hearing about the monetary rewards that other firms have achieved through Six Sigma methodology and are eager to cash in on similar benefits. The book *Implementing Six Sigma: Smarter Solutions using Statistical Methods* (Wiley, 1999), by Forrest W. Breyfogle III, was written as a practical guide to help organizations in both industry and academia with the *wise* implementation and orchestration of Six Sigma techniques.

Implementing Six Sigma has been a great success; however, many people have expressed the desire for a book that will help management decide if they should implement Six Sigma and then guide them through the process. To fulfill this need, we wrote *Managing Six Sigma*.

The purpose of this book is to build awareness of the wise application of Six Sigma tools and how they can be important in the "big picture." Because of this focus, there may be references to tools with which the reader is not familiar. To help the reader gain a basic understanding of unfamiliar terms, we provide in the glossary a brief description of most Six Sigma tools discussed in the text. *Implementing Six Sigma* can be consulted for more details about individual tools or applications (see the outline of that book below).

Six Sigma can improve the bottom line of an organization—if implemented *wisely*. An organization can get more with less using Six Sigma; for example, it can use fewer runs and samples and obtain more information. However, if the techniques are not used wisely, there is a considerable danger that the program will be counterproductive and frustrating. Organizations can sometimes get so involved in how to count and report defects that they lose sight of the real value of Six Sigma—orchestrating process improvement and reengineering in such a way that they achieve significant bottom-line benefits through the implementation of statistical techniques. Six Sigma efforts need to be orchestrated toward achieving Smarter Solutions (Smarter Solutions, Smarter Six Sigma Solutions, and S^4 are service marks belonging to Forrest W. Breyfogle III).

If an organization does not apply Six Sigma techniques wisely, the methodology will fail. When this occurs, there is a tendency to believe that the statistical techniques are not useful, when in fact the real problem is how Six Sigma as a program was implemented or how individual techniques were effectively applied.

This book uses the term "Smarter Six Sigma Solutions (S^4)" to describe our implementation strategy for Six Sigma. Another description for this S^4 activity is "$marter Six Sigma $olutions" assessments, since a major focus is determining that the right measurements and actions are being taken relative to bottom-line benefits. With S^4 activities an environment is created where there is knowledge-centered activity (KCA) focus. KCA describes efforts for wisely obtaining knowledge and/or utilizing the knowledge of organizations and processes.

GE and other companies have used the terms "Black Belts" and "Green Belts" (Cheek, 1992; GE, 1996; GE, 1997; Lowe, 1998; Slater, 1999; Pyzdek, 1999; Harry, 2000) to describe people who actively apply Six Sigma techniques. These people may be assigned to this role either full-time or part-time. in this book we will use the terms "Black Belt" and "Green Belt" along with the term "Six Sigma Practitioner" to describe people who implement Six Sigma techniques.

This book refers the reader to Forrest W. Breyfogle III's *Implementing Six Sigma: Smarter Solutions Using Statistical Methods*, for more descriptions of tools, implementation techniques, and strategies. References to the earlier work appear as Breyfogle (1999), *Implementing Six Sigma,* which contains the following sections:

- Phase 0: S^4 Deployment
- Phase 1: S^4 Measurement
- Phase 2: S^4 Analysis
- Phase 3: S^4 Improvement
- Phase 4: S^4 Control

Implementing Six Sigma contains more than 100 examples within the following chapters:

Phase 0: S^4 Deployment Strategy Phase

1. Six Sigma Overview and Implementation
2. Knowledge Centered Activity (KCA) and Process Improvement

Phase I: S⁴ Measurement Phase

3. Overview of Descriptive Statistics and Experimentation Traps
4. Process Flowcharting
5. Basic Tools
6. Probability
7. Overview of Distributions and Statistical Processes
8. Probability and Hazard Plotting
9. Six Sigma Measurements
10. Basic Control Charts
11. Process Capability and Process Performance
12. Measurement Systems Analysis (Gage R&R)
13. Cause-and-Effect Matrix and Quality Function Deployment (QFD)
14. Failure Mode and Effects Analysis (FMEA)

Phase II: S⁴ Analysis Phase

15. Visualization of Data
16. Confidence Intervals and Hypothesis Tests
17. Inferences: Continuous Response
18. Inferences: Attribute Response
19. Comparison Tests: Continuous Response
20. Comparison Tests: Attribute Response
21. Bootstrapping
22. Variance Components Analysis
23. Correlation and Simple Linear Regression
24. Single-factor (one-way) Analysis of Variance
25. Two-factor (two-way) Analysis of Variance
26. Multiple Regression

Phase III: S⁴ Improvement Phase

27. Benefiting from Design of Experiments
28. Understanding the Creation of Full and Fractional Factorial 2^k DOEs
29. Planning 2^k DOEs
30. Design and Analysis of 2^k DOEs
31. Other DOE Considerations
32. Variability Reduction Through DOE and Taguchi Considerations
33. Response Surface Methodology

Phase IV: S⁴ Control Phase

34. Short-run and Target Control Charts
35. Other Control Charting Alternatives
36. Exponentially Weighted Moving Average (EWMA) and Engineering Process Control (EPC)
37. Pre-Control Charts
38. Control Plan and Other Strategies
39. Reliability Testing/Assessment: Overview
40. Reliability Testing/Assessment: Repairable System
41. Reliability Testing/Assessment: Nonrepairable Devices
42. Pass/Fail Functional Testing
43. Application Examples

To meet the needs of a diverse readership, we adhered to the following guidelines in the writing of this book:

- Most chapters and sections are small, descriptive, and contain many illustrations. The table of contents can be very useful to quickly locate techniques and examples helpful in solving a particular problem.
- Equations and formulas are presented only when we believe they are absolute necessary to describe a methodology.
- The symbols and glossary sections are intended as a handy reference that provides fuller explanations whenever a concise definition or an unfamiliar statistical term or symbol is encountered in the text.
- The focus of the book is on manufacturing, development, and service/transactional examples that serve as a bridge between Six Sigma techniques and a variety of real-world situations.
- The details of implementing Six Sigma are not included in this book, since these techniques are described in Breyfogle (1999), *Implementing Six Sigma.*
- Even though Breyfogle (1999), *Implementing Six Sigma,* includes specific information for those undertaking the details of implementing Six Sigma, managers should find the S⁴ assessments sections very helpful. These sections, found at the end of most chapters, provide guidance inthe effective orchestration of Six Sigma implementation.

ACKNOWLEDGMENTS

The authors are appreciative of those who helped us define major topics for this book and develop a structure in which to present them. They were Paul Iglesias, Mike Kirchoff, John W. Knickel, Alejandra Ajuria Martin, Terry J. O'Connell, and Roland Schimpke. The authors are also appreciative of Titanium Metals Corporation (TIMET) for letting us use information from one of their projects and Mark Wallace for his input to an early manuscript.

The terms key process output variable (KPOV), key process input variable (KPIV), and the term used to describe a simplified quality function deployment (QFD) matrix—cause-and-effect matrix—were coined by Sigma Breakthrough Technologies, Inc. (SBTI).

CONTACTING THE AUTHORS

Your comments and suggestions will be considered as we prepare future editions of this book, since we work at practicing what we preach. Also, we conduct both public and in-house Six Sigma workshops from this book and Breyfogle (1999), *Implementing Six Sigma*, utilizing S^4 techniques. Contact us if you would like information about these workshops. We can be reached through the following e-mail address: forrest@smartersolutions.com. You might also find useful the articles and additional implementation ideas at www.smartersolutions.com.

Please send your Table 8.1 survey responses to Smarter Solutions Inc., 1779 Wells Branch Parkway, #110B-281, Austin, TX, 78728, USA

PART 1

WHY SIX SIGMA

1

HOW SIX SIGMA COMPARES TO OTHER QUALITY INITIATIVES

As the competition gets tougher, there is more pressure on organizations to improve quality and customer satisfaction while decreasing costs and increasing work output. This becomes an increasingly difficult challenge when there are fewer resources available. Peter Senge (1990) writes, "Learning disabilities are tragic in children, but they are fatal in organizations. Because of them, few corporations live even half as long as a person—most die before they reach the age of forty." "Learning organizations" defy these odds and overcome learning disabilities to understand threats and recognize new opportunities. Six Sigma can help organizations learn and excel at the challenges they encounter—if it is implemented *wisely*.

A question we frequently hear from executives is "How does Six Sigma fit with other company initiatives?" We believe that Six Sigma should not be considered just another initiative but should integrate other programs and initiatives at a higher level as part of an overall business strategy. Six Sigma should not replace other initiatives, but instead offer a tactical methodology to determine the best approach for a given situation/process.

Our Smarter Six Sigma Solutions (S⁴) business strategy offers a road map for changing data into knowledge that leads to new opportunities. The major components to consider during Six Sigma implementation are "metrics" and "strategy," as shown in Figure 1.1. The upper half of the figure involves the measurement of how well business processes meet their goals. The success of Six Sigma is linked to a set of cross-functional metrics that lead to significant improvements in customer satisfaction and bottom-line benefits. Organizations

3

4

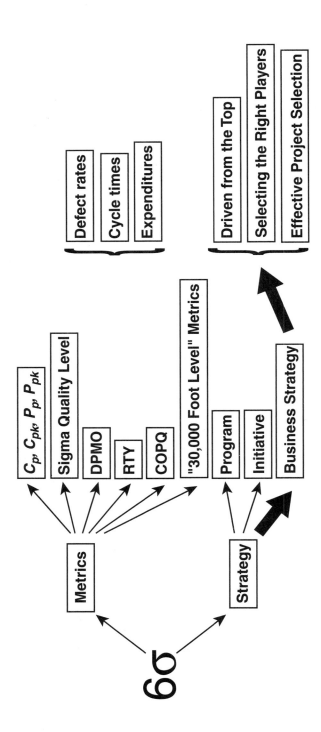

Figure 1.1 Six Sigma implementation considerations

do not necessarily need to use all the measurements listed (often presented within typical Six Sigma programs). It is most important to choose the best set of metrics for a situation, metrics that yield insight into a situation or process.

Our S^4 approach advocates the development of cross-functional teams to provide a holistic approach to problem solving, encompassing all levels of complex processes. We often describe the methodology as a murder mystery, where practitioners are determining "who done it?" or, equivalently, "what is the major cause of defects in a process?" By following a structured methodology, project teams can determine the "biggest hitters" and make substantial areas for improvement that provide real benefits to an organizations bottom line.

Subsequent chapters of this book will provide the details of effective Six Sigma metrics and the importance of implementing Six Sigma as a business strategy. In this chapter, we first discuss current myths surrounding Six Sigma. We then provide a brief history of quality leaders and other quality systems that preceded the creation of Six Sigma. Last, we answer a few of the more frequently asked questions (FAQs) about Six Sigma.

1.1   WHAT IS SIX SIGMA?

Some people view Six Sigma quality as merely a rigorous application of basic and advanced statistical tools throughout an organization. There are a number of Six Sigma consultants and training organizations that have simply repackaged the statistical components of their previous TQM programs and renamed them "Six Sigma." These groups would define Six Sigma quality in terms like those in the upper half of Figure 1.1.

Others view Six Sigma as merely a sophisticated version of Total Quality Management (TQM), as represented by the lower half of Figure 1.1. They see it as an advanced form of TQM in which various continuous improvement systems must be put in place with a small amount of statistical analyses added in for good measure.

The S^4 view of Six Sigma emphasizes an intelligent blending of the wisdom of the organization with proven statistical tools to improve both the efficiency and effectiveness of the organization in meeting customer needs. The ultimate goal is not improvement for improvement's sake, but rather the creation of economic wealth for the customer and provider alike. Our Smarter Solutions approach recommends that Six Sigma be viewed as a strategic business initiative rather than a quality program. This implies, not that Six Sigma replaces existing and ongoing quality initiatives in an organization, but that senior management focuses on those processes identified as critical-to-quality in the eyes of customers. Those critical systems are then the subject of intense scrutiny and improvement efforts, using the most powerful soft and hard skills the organization can bring to bear. The success of each and every Six Sigma initiative is

linked to a set of multidimensional metrics that demand world-class performance, which, if achieved, lead to significant improvements in market share,
new product development, customer satisfaction, and shareholder wealth.

Later in this chapter and in subsequent chapters, we will spend more time
explaining what Six Sigma is. But first we will explain what it is not. We will
start by dispelling the 10 myths of Six Sigma (Snee, 1999b) listed in Table 1.1.

TABLE 1.1   The 10 Myths of Six Sigma

Six Sigma Myths

- Works only in manufacturing
- Ignores the customer in search of bottom-line benefits
- Creates a parallel organization
- Is an add-on effort
- Requires massive training
- Requires large teams
- Creates bureaucracy
- Is just another quality program
- Requires complicated, difficult statistics
- Is not cost-effective

Works only in manufacturing

Much of the initial success in applying Six Sigma was based on manufacturing
applications; however, recent publications have addressed other applications of
Six Sigma. Breyfogle (1999), *Implementing Six Sigma*, includes many transac-
tional/service applications. In GE's *1997 Annual Report* (GE, 1997), CEO Jack
Welch proudly states that Six Sigma "focuses on moving every process that
touches our customers—every product and *service* [emphasis added]—toward
near-perfect quality."

Ignores the customer in search of profits

This statement is not myth, but rather misinterpretation. Projects worthy of Six
Sigma investment should (1) be of primary concern to the customer, and (2) have
the potential for significantly improving the bottom line. Both criteria must be
met. The customer is driving this boat. In today's competitive environment,
there is no surer way of going out of business than to ignore the customer in a
blind search for profits.

Creates a parallel organization

An objective of Six Sigma is to eliminate every ounce of organizational waste that can be found and then reinvest a small percentage of those savings to continue priming the pump for improvements. With the large amount of downsizing that has taken place throughout the world during the past decade, there is no room or inclination to waste money through the duplication of functions. Many functions are understaffed as it is. Six Sigma is about nurturing any function that adds significant value to the customer while adding significant revenue to the bottom line.

Requires massive training

"Valuable innovations are the positive result of this age [we live in], but the cost is likely to be continuing system disturbances owing to members' nonstop tinkering. . . . [P]ermanent white water conditions are regularly taking us all out of our comfort zones and asking things of us that we never imagined would be required. . . . It is well for us to pause and think carefully about the idea of being continually catapulted back into the beginner mode, for that is the real meaning of being a continual learner. . . . We do not need competency skills for this life. We need incompetency skills, the skills of being effective beginners." (Vaill)

Is an add-on effort

This is simply the myth "creates a parallel organization" in disguise. Same question, same response.

Requires large teams

There are many books and articles within business literature declaring that teams have to be small if they are to be effective. If teams are too large, the thinking goes, a combinational explosion occurs in the number of possible communication channels between team members, and hence no one knows what the other person is doing.

Creates bureaucracy

A dictionary definition of bureaucracy is "rigid adherence to administrative routine." The only thing rigid about wisely applied Six Sigma methodology is its relentless insistence that customer needs be addressed.

Is just another quality program

Based upon the poor performance of untold quality programs during the past three to five decades (Micklethwait and Wooldridge, 1997), an effective quality program would be welcome. More to the point (Pyzdek, 1999c), Six Sigma is "an entirely new way to manage an organization."

Requires complicated, difficult statistics

There is no question that a number of advanced statistical tools are extremely valuable in identifying and solving process problems. We believe that practitioners need to possess an analytical background and understand the wise use of these tools, but do not need to understand all the mathematics behind the statistical techniques. The wise application of statistical techniques can be accomplished through the use of statistical analysis software.

Is not cost-effective

If Six Sigma is implemented wisely, organizations can obtain a very high rate of return on their investment within the first year.

1.2   THE QUALITY LEADERS

Numerous individuals have had a significant impact on the Six Sigma quality movement in the United States and abroad. Unfortunately, we don't have time or space to acknowledge them all. The six we have chosen to give a diverse perspective are W. Edwards Deming, William Conway, Joseph Juran, Philip Crosby, Genichi Taguchi, and Shigeo Shingo. This brief review is more than a historical perspective. At the conclusion of this section, we will discuss how Six Sigma relates to the previous generation of quality programs.

W. Edwards Deming

Dr. W. Edwards Deming was born on October 14, 1900. By 1928, he had earned a B.S. in engineering at the University of Wyoming in Laramie, an M.S. in mathematics and physics at the University of Colorado, and a Ph.D. in mathematical physics at Yale. He is best known for revitalizing Japanese industry after the end of World War II. In 1950, he visited Japan at the request of the Secretary of War to conduct a population census. While there he was invited by a member of the Japanese Union of Science and Engineering (JUSE) to lecture on statistical methods for business at a session sponsored by the Keidanren, a prestigious society of Japanese executives. Deming told the Japanese leaders that they could "take over the world" if they followed his teachings. The rest is history. Today the the most prestigious quality award in Japan is called the "Deming Prize." Deming, who has been heralded as the "founder of the Third Wave of the Industrial Revolution," often sounded like a quality crusader when he issued statements such as "It is time to adopt a new religion in America." He gained national recognition in 1980 when he was interviewed for the NBC television "white paper," "If Japan Can, Why Can't We?"

Dr. Deming is equally well known for his "Fourteen Points" (Table 1.2) and "Seven Deadly Diseases" (Table 1.3) (Deming, 1986). He made numerous modifications to the first list throughout his life as he grew in knowledge and wisdom. Additional references that provide commentary on Deming's Fourteen Points are provided in the references section of this book (Breyfogle, 1999; Neave, 1990; Scherkenbach, 1988; Gitlow and Gitlow, 1987).

TABLE 1.2   Deming's Fourteen Points

1. Create constancy of purpose toward improvement of product and service, with the aim to become competitive and stay in business and to provide jobs.
2. Adopt the new philosophy. We are in a new economic age. Western management must awaken to the challenge, must learn their responsibilities, and take on leadership for change.
3. Cease dependence on inspection to achieve quality. Eliminate the need for inspection on a mass basis by building quality into the product in the first place.
4. End the practice of awarding business on the basis of price tag. Instead, minimize total cost. Move toward a single supplier for any one item, on a long-term relationship of loyalty and trust.
5. Improve constantly and forever the system of production and service, to improve quality and productivity, and thus constantly decrease costs.
6. Institute training on the job.
7. Institute leadership. The aim of supervision should be to help people and machines and gadgets to do a better job. Supervision of management is in need of overhaul, as well as supervision of production workers.
8. Drive out fear, so that everyone may work effectively for the company.
9. Break down barriers between departments. People in research, design, sales, and production must work as a team to foresee problems of production and in use that may be encountered with the product or service.
10. Eliminate slogans, exhortations, and targets for the work force asking for zero defects and new levels of productivity. Such exhortations only create adversary relationships, as the bulk of the causes of low quality and low productivity belong to the system and thus lie beyond the power of the work force.
11a. Eliminate work standards (quotas) on the factory floor. Substitute leadership.
11b. Eliminate management by objective. Eliminate management by numbers, numerical goals. Substitute leadership.

12a. Remove barriers that rob the hourly worker(s) of their right to pride of workmanship. The responsibility of supervisors must be changed from sheer numbers to quality.

12b. Remove barriers that rob people in management and in engineering of their right to pride of workmanship. This means, inter alia, abolishment of the annual or merit rating and of managing by objective.

13. Institute a vigorous program of education and self-improvement.

14. Put everybody in the company to work to accomplish the transformation. The transformation is everybody's job.

TABLE 1.3   Deming's Seven Deadly Diseases

1. Lack of constancy of purpose to plan product and service that will have a market and keep the company in business, and provide jobs.

2. Emphasis on short-term profits: short-term thinking (just the opposite from constancy of purpose to stay in business), fed by fear of unfriendly takeover, and by push from bankers and owners for dividends.

3. Evaluation of performance, merit rating, or annual review.

4. Mobility of management: job hopping.

5. Management by use only of visible figures, with little or no consideration of figures that are unknown or unknowable.

6. Excessive medical costs.

7. Excessive costs of liability, swelled by lawyers that work on contingency fees.

Deming's basic quality philosophy is that productivity improves as variability decreases. Since all things vary, he says, statistical methods are needed to control quality. "Statistical control does not imply absence of defective items. It is a state of random variation, in which the limits of variation are predictable," he explains.

There are two types of variation: chance and assignable. Deming states that "the difference between these is one of the most difficult things to comprehend." His "red bead experiment" (Walton, 1986) revealed the confusion generated by not appreciating the difference between the two. It is a waste of time and money to look for the cause of chance variation, yet, he says, this is exactly what many companies do when they attempt to solve quality problems without using statistical methods. He advocates the use of statistics to measure performance in all areas, not just conformance to product specifications. Furthermore, he says, it is not enough to meet specifications; one has to keep working to reduce the variation as well.

Deming is extremely critical of the U.S. approach to business management and is an advocate of worker participation in decision making. He claims that management is responsible for 94% of quality problems, and he believes that it is management's task to help people work smarter, not harder. There is much in common between Dr. Deming's teachings and the Smarter Six Sigma Solutions (S^4) methodology.

William E. Conway

William E. Conway (Conway, 1999) is the founder, chairman, and CEO of Conway Management Company. He attended Harvard College and is a graduate of the United States Naval Academy.

In 1980, after viewing the NBC television broadcast of "If Japan Can, Why Can't We?", he invited Dr. W. Edwards Deming to Nashua Corporation, becoming one of the first American executives to approach Deming for help. The visits went on for three years. Because of his close and early association with Deming, Conway is sometimes described as a "Deming disciple," but he has developed his own philosophy of and approach to quality.

Conway has created a system of management that lets organizations achieve lasting, bottom-line improvements. That system is called "The Right Way to Manage." Much of the content of this two-day course is also presented in one of Mr. Conway's quality books (Conway, 1992). In this book he discusses the impact of variation on quality, and introduces his approach to using the seven simple tools (including control charts) to eliminate waste. The core activity of the Conway System is eliminating waste in all processes. While most executives are familiar with the waste associated with manufacturing operations, it actually exists throughout all functions of an organization. This approach has been adopted by diverse organizations in a wide range of industries: wholesale, retail, service, manufacturing, distribution, health care, and government.

Conway notes that it takes more than just a critical eye to spot the hidden or "institutionalized" waste. Training and education are needed, as well as the right tools and techniques. He reports a solid return on investment (ROI) in his firm's education and training programs; ROIs of 4:1 to 10:1 are not uncommon.

In his most recent book (Conway, 1994), Mr. Conway delves into some of the infrastructure and implementation issues involved in eliminating organizational waste across the board.

Joseph M. Juran

Joseph Moses Juran (Juran, 1999a) was born December 24, 1904, in Braila, Romania. In 1920, Juran enrolled at the University of Minnesota. In 1924, he graduated with a B.S. degree in electrical engineering and took a job with Western Electric in the Inspection Department of the Hawthorne Works in Chicago. In 1926, a team from Bell Laboratories (including Walter Shewhart and Harold Dodge) made a visit to the factory with the intention of applying the laboratory tools and methods they had been developing. Juran was selected as one of 20 trainees, and he was subsequently chosen as one of two engineers to work in the newly created Inspection Statistical Department. In 1928, Juran authored his first work on quality, a pamphlet that became an input to the well-known *AT&T Statistical Quality Control Handbook*, which is still published today.

In 1937, Juran conceptualized the Pareto principle. In December 1941, he took a leave of absence from Western Electric to serve in Washington with the Lend-Lease Administration. It was here that he first experimented with what today might be called "business process reengineering." His team successfully eliminated the paper logjam that kept critical shipments stalled on the docks. In 1945, Juran left Washington and Western Electric to work as an independent consultant. In 1951, his standard reference work on quality control, *Quality Control Handbook* (Juran, 1999b), was first published.

In 1954, the Union of Japanese Scientists and Engineers (JUSE) and the Keidanren invited the celebrated author to Japan to deliver a series of lectures. These talks addressed managing for quality, and were delivered soon after Dr. W. Edwards Deming had delivered his lectures on statistical quality methods.

Dr. Juran has been called the father of quality, a quality guru, and the man who "taught quality to the Japanese." He is recognized as the person who added the "human dimension" to quality, expanding it beyond its historical statistical origins to what we now call total quality management. Says Peter Drucker, "Whatever advances American manufacturing has made in the last thirty to forty years, we owe to Joe Juran and to his ... work." A few of Juran's important quality books are listed in the reference section of this book (Juran, 1964, 1988, 1989, 1992).

Philip B. Crosby

Phil Crosby (Crosby, 1999) was born in Wheeling, West Virginia, on June 18, 1926. He started work as a quality professional on an assembly line at Crosley in 1952 after serving in World War II and Korea. He took it upon himself to try to convince management that it was more profitable to prevent problems than to fix them. He worked for Crosley from 1952 to 1955; Martin-Marietta from 1957 to 1965; and ITT from 1965 to 1979. As quality manager for Martin-Marietta he created the Zero Defects concept and program. During his 14 years as corporate vice president for ITT, he worked with many industrial and service companies around the world, implementing his pragmatic philosophy.

In 1979, Crosby founded Philip Crosby Associates, Inc. (PCA), and over the next 10 years grew it into a publicly traded organization with 300 employees around the world. In 1991, he retired from PCA and founded Career IV, Inc., a company that provided lectures and seminars aimed at helping current and prospective executives grow. In 1997 he purchased the assets of PCA and established Philip Crosby Associates II, Inc. Now his "Quality College" operates in more than 20 countries around the world.

Philip Crosby's lectures provide a thoughtful and stimulating discussion of managements' role in causing their organizations, their employees, their suppliers, and themselves to be successful. He has published 13 books, all of which have been best-sellers. His first business book, *Quality Is Free* (Crosby, 1979), has been credited with beginning the quality revolution in the United States and Europe. A few of his other titles are included in the references section of this book (Crosby, 1984, 1992).

Philip Crosby's books are easy to read and explain his program for quality improvement and defect prevention. He emphasizes the importance of management's role in making quality happen. His approach is team based but is not highly statistical, as are some of the competing approaches to TQM. He does teach the use of the simple tools and statistical quality control, but in a very generic, non-mathematical sense. He is an interesting and knowledgeable speaker who both entertains and educates his audiences. He has certainly had an impact on the quality revolution over the years.

Dr. Genichi Taguchi

Dr. Genichi Taguchi's (Taguchi, 1999) system of quality engineering is one of the great engineering achievements of the twentieth century. He is widely acknowledged as a leader in the U.S. industrial quality movement. His philosophy began taking shape in the early 1950s when he was recruited to help correct postwar Japan's crippled telephone system. After noting the deficiencies inherent in the trial-and-error approach to identifying problems, he developed his own integrated methodology for designed experiments.

Systematic and widespread application of Dr. Taguchi's philosophy, and of his comprehensive set of experimental design decision-making tools, has contributed significantly to Japan's prowess in rapidly producing world-class, low-cost products.

The experimental procedures developed and taught by Genichi Taguchi (Taguchi and Konishi, 1987; Ross, 1988) have met with both skepticism and acclaim. Some nonstatisticians find his techniques highly practical and useful. Most statisticians, however, have identified and published evidence that use of some of his techniques can lead to erroneous conclusions.

We avoid these and other controversies, and instead make use of Taguchi Methods where they are of benefit and where they avoid known pitfalls and weaknesses. The two major areas where Dr. Taguchi's contributions to quality are recognized are (1) variance-reduction strategies and (2) robust design techniques.

Shigeo Shingo

Shigeo Shingo was born in Saga City, Japan, in 1909. In 1930, he graduated with a degree in mechanical engineering from Yamanashi Technical College and went to work for the Taipei Railway Factory in Taiwan. Subsequently he became a professional management consultant in 1945 with the Japan Management Association. It was in his role as head of the Education Department in 1951, that he first heard of, and applied, statistical quality control. By 1954 he had studied 300 companies. In 1955, he took charge of industrial engineering and factory improvement training at the Toyota Motor Company for both its employees and parts suppliers (100 companies). During the period 1956–1958, at Mitsubishi Heavy Industries in Nagasaki, Shigeo was responsible for reducing the time for hull assembly of 65,000-ton supertankers from four months to two months. In 1959, he left the Japan Management Association and established the Institute of Management Improvement, with himself as president. In 1962, he started industrial engineering and plant-improvement training at Matsushita Electrical Industrial Company, where he trained some 7,000 people.

Shingo's supreme contribution in the area of quality was his development in the 1960s of poka-yoke (pronounced "POH-kah YOH-kay"). The term comes from the Japanese words "poka" (inadvertent mistake) and "yoke" (prevent). The essential idea of poka-yoke is to design processes so mistakes are impossible to make or at least easily detected and corrected.

Poka-yoke devices fall into two major categories: prevention and detection. A prevention device affects the process in such a way that it is impossible to make a mistake. A detection device signals the user when a mistake has been made, so that the user can quickly correct the problem. Shingo's first poka-yoke device was a detection implement he created after visiting Yamada Electric in 1961. He was told of a problem in the factory that occasionally led to workers' assembling a small, two-push-button switch without inserting a spring under each push-button. The problem of the missing spring was both costly and embarrassing, but despite everyone's best effort the problem continued. Shingo's solution was simple. The new procedure completely eliminated the problem:

- Old method: a worker began by taking two springs out of a large parts box and then assembled a switch.
- New approach: a small dish was placed in front of the parts box and the worker's first task was to take two springs out of the box and place them on the dish. Then the worker assembled the switch. If any spring remained on the dish, then the worker knew that he or she had forgotten to insert it.

Although Shigeo Shingo is perhaps less well known in the West, his impact on Japanese industry has been immense. Norman Bodek, president of Productivity, Inc., stated: "If I could give a Nobel Prize for exceptional contributions to world economy, prosperity, and productivity, I wouldn't have much difficulty selecting a winner—Shigeo Shingo." Shingo died in November of 1990 at 81 years of age.

1.3 TQM AND SIX SIGMA

The six quality professionals we discussed have a lot in common with respect to the intellectual content of their quality approaches. They have chosen to emphasize different aspects of the body of knowledge known as quality. Deming and Taguchi emphasize the statistical aspects of quality; Deming, Crosby, and Juran focus on management's role in the improvement process; Conway relentlessly drives home the need to eliminate waste wherever it can be found; Shingo preaches error-proofing. There is no doubt that these individuals had a marked impact on how we thought about quality during the second half of the twentieth century. But does their expertise transfer to bottom-line benefits for organizations? In a survey of chief executive officers who subscribe to *Electronic Business,* only 16% reported improved market share and only 13% reported improved operating income or profits as a result of their quality efforts (Boyett et al., 1992). Mikel J. Harry (2000b) reports that even as increasing numbers of companies jumped on the quality "express" during this time, fewer were reporting any meaningful impact on profitability.

A study published by the American Society for Quality (Bergquist and Ramsing, 1999) compares the performance of winners and applicants of both the Malcolm Baldrige National Quality Award (MBNQA) and state-level award programs with nonapplicants in the same two categories. The authors are unable to "conclusively determine whether quality award winning companies perform better than others [i.e., nonapplicants]."

One of the problems that plagued many of the early TQM initiatives was the preeminence placed on quality at the expense of all other aspects of the business. Some organizations went bankrupt, or experienced severe financial consequences, in the rush to make quality "first among equals." Harry (2000a) notes that "what's good for the customer is not always good for the provider." The disconnect between management systems designed to measure customer satisfaction and those designed to measure provider profitability often led to unwise investments in quality.

We believe that Six Sigma is more than a simple repackaging of the best from other TQM programs. Tom Pyzdek (1999c) agrees: "Six Sigma is such a drastic extension of the old idea of statistical control as to be an entirely different subject. . . . In short, Six Sigma is . . . an entirely new way to manage an organization. . . . Six Sigma is not primarily a technical program; it's a management program."

Ronald Snee (1999a) points out that although some people believe it is nothing new, Six Sigma is unique in its approach and deployment. He defines Six Sigma as a strategic business improvement approach that seeks to increase *both* customer satisfaction and an organization's financial health. Snee goes on to claim that eight characteristics account for Six Sigma's increasing bottom-line success and popularity with executives:

- Bottom-line results expected and delivered
- Senior management leadership
- A disciplined approach (Measure, Analyze, Improve, and Control)
- Rapid (3–6 month) project completion
- Clearly defined measures of success
- Infrastructure roles for Six Sigma practitioners and leadership
- Focus on customers and processes
- A sound statistical approach to improvement

Other quality initiatives have laid claim to one or two of these characteristics, but only Six Sigma attributes its success to the simultaneous application of all eight.

Mikel Harry (2000b) goes even farther and claims that Six Sigma represents a new, holistic, multidimensional systems approach to quality that replaces the "form, fit and function specifications" of the past. In his New World view of quality, both the customer and provider receive benefits of *utility, availability,* and *worth.*

- Utility
  - Form (is in a pleasing/effective physical form)
  - Fit (made to the correct size/dimensions)
  - Function (performs all required tasks well)
- Availability
  - Volume (production capacity, inventory levels, and/or distribution channels are adequate to meet varying demand)
  - Timing (a short product development cycle time; lot size = 1; and/or JIT availability)
- Worth
  - Intellectual (innovative; sought by early adopters)
  - Emotional (satisfies one's pride, passions, or psyche)
  - Economic (customer perceived value [quality-to-price ratio] is high)

We believe Six Sigma encompasses all of these things and more. As Figure 1.1 depicts, the S^4 approach encourages the following:

- Creating an infrastructure with executive management that supports cultural change
- Using "30,000 foot level" metrics and the cost of poor quality when selecting projects (to be discussed in detail in later chapters)
- Selecting the "right" people and realigning resources as needed
- Selecting the right provider with an effective training approach
- Selecting the right projects that meet customer needs and provide bottom-line benefits
- Developing project metrics that yield insight into the process and discourage the "firefighting" approach to solving problems
- Applying the right tools at the right time
- Setting SMART goals (Simple, Measurable, Agreed to, Realistic, Time-based)
- Making holistic process improvements through wisdom of the organization

1.4 SIX SIGMA AND INTEGRATION OF THE STATISTICAL TOOLS

The Smarter Six Sigma Solutions (S^4) approach involves identifying projects that target the customer's concerns and have the potential for significant payback. This is accomplished using a team-based approach (demanding soft skills) coupled with a wide range of statistical tools (demanding hard skills). Zinkgraf and Snee (1999) have drawn up a list of the eight tools they believe are essential to getting the job done (Table 1.4). These tools represent a subset of a much larger array of statistical and management tools at one's disposal. Breyfogle (1999), in *Implementing Six Sigma*, illustrates a 21-step integration of the tools, which is shown in Table 1.5 and further elaborated upon in later chapters.

TABLE 1.4   The "Right" Tools

Sigma Breakthrough Technologies Inc. (SBTI) "Right" Tools

Process Maps

Cause-and-Effects Matrix

Measurement Systems Analysis (Gage R&R)

Process Capability Studies

Failure Modes and Effects Analysis (FMEA)

Multi-Vari Studies

Design of Experiments (DOE)

Control Plans

TABLE 1.5   The 21-Step Integration of the Tools [From Breyfogle (1999), with permission]

Step	Action	Participants	Source of Information
1	Identify critical customer requirements from a high-level project measurement point of view. Identify KPOVs that will be used for project metrics. Implement a balanced scorecard considering **COPQ** and **RTY** metrics.	S⁴ black belt and champion	Current data
2	Identify team of key "stakeholders."	S⁴ black belt and champion	Current data
3	Describe business impact. Address financial measurement issues of project.	S⁴ black belt and finance	Current data
4	Plan overall project.	Team	Current data
5	Start compiling project metrics in time series format. Utilize a sampling frequency that reflects "long-term" variability. Create **run charts** and **control charts** of KPOVs.	Team	Current and collected data
6	Determine "long-term" **process capability/ performance** of KPOVs. Quantify nonconformance proportion. Determine baseline performance. **Pareto chart** types of defects.	Team	Current and collected data
7	Create a process **flowchart/ process map.**	Team	Organization wisdom
8	Create a **cause and effect diagram** to identify variables that can affect the process output.	Team	Organization wisdom

9	Create a **cause and effect matrix** assessing strength of relationships that are thought to exist between KPIVs and KPOVs.	Team	Organization wisdom
10	Conduct a **measurement systems analysis.** Consider a **variance components** analysis.	Team	Collected data
11	Rank importance of KPIVs using a **Pareto chart.**	Team	Organization wisdom
12	Prepare a focused **FMEA.** Assess **current control plans**.	Team	Organization wisdom
13	Collect data for assessing the KPIV/KPOV relationships that are thought to exist.	Team	Collected data
14	Create **multi-vari charts** and **box plots.**	Team	Passive data analysis
15	Conduct **correlation** studies.	Team	Passive data analysis
16	Assess statistical significance of relationships using **hypothesis tests.**	Team	Passive data analysis
17	Conduct **regression** and **analysis of variance (ANOVA)** studies.	Team	Passive data analysis
18	Conduct designed experiments **(DOE)** and **response surface methods (RSM)** analyses.	Team	Active experimentation
19	Determine optimum operating windows of KPIVs from **DOEs, response surface methods (RSM)** and other tools.	Team	Passive data analysis and active experimentation

| 20 | Update **control plan.** Implement **control charts** to timely identify special cause excursions of KPIVs. | Team | Passive data analysis and active experimentation |
| 21 | Verify process improvements, stability, and **capability/ performance** using demonstration runs. | Team | Active experimentation |

In addition to the array of tools, there are numerous implementation strategies for introducing these tools into S^4 projects depending on whether the application involves manufacturing, development, or service/transactional processes. The order in which these tools are introduced is also very important. Part 1 of this book gives a concise, generic road map to the essential tools that may be utilized for any project. Part 4, "Applying Six Sigma," includes other possible sequences for approaching an S^4 project and integrating tools appropriately for manufacturing, service/transactional, and development applications.

There are no fixed rules that specify the correct order of tool usage. Each project can evolve in its own unique way, demanding its own order of tool usage. The role of the Six Sigma practitioner is to determine tool selection and their use. As organizations acquire experience using Six Sigma tools and techniques, team members can become as capable as formally trained Black Belts in utilizing the tools of Six Sigma.

The first 17 steps listed in Table 1.5 primarily involve collecting and analyzing process output data, compiling information on what can be done differently to make improvements using the wisdom of the organization, and analyzing data passively in an attempt to detect a cause-and-effect relationship between input and output variables. We are suggesting that the first step in an S^4 approach is not to immediately conduct a designed experiment on the production floor, but rather to identify the requirements deemed critical by the customer. Notice that more advanced statistical tools are found near the middle and end of the 21-step process. Examples of these include regression analysis, hypothesis testing, analysis of variance, variance components analysis, multi-vari charts, measurement systems analysis (Gage R&R), design of experiments (DOE), response surface methods (RSM), statistical process control (SPC), and process capability/performance analyses.

It is important to extract and analyze the existing information on a process before proceeding to change it. We suggest the early use of high-level control charts, which we call "30,000-foot level" control charts. These charts can be useful in establishing a baseline of the current process from which monetary and other project parameters can be quantified. Later in this book, we describe how this chart can be used to get organizations out of the "firefighting mode." The simple tools of Pareto charts, flowcharts, process maps, and so forth also come in handy at the beginning of a project as ways of displaying and analyzing data. It is important to realize that the assigned Six Sigma practitioners are involved in every step of the process. They are an integral part of the project team but are identified individually in the first three steps, since prior to step 4 the team itself does not yet exist. There is important work to be done before team members are identified and trained.

1.5   LEAN MANUFACTURING AND SIX SIGMA

Most of the information in this section was reproduced (with minor additions, deletions, and modifications) from an article by Denecke (1998), and is reproduced with the permission of Honeywell International Inc. (formerly AlliedSignal). We decided to share the article with readers of this book. We have paraphrased it extensively, with permission, and included our own response to the article.

It should be noted that in our opinion the differences between how Six Sigma and lean manufacturing would address a project are even less than what is suggested by this article. We believe that Six Sigma should have metrics, such as cycle time, that lead to the application of the discipline of lean manufacturing when it is most appropriate. In addition, the techniques of Six Sigma should be used within processes to reduce defects, which can be a very important prerequisite for a lean manufacturing project to be successful.

Background

Currently there are two premier approaches to improving manufacturing operations. One is lean manufacturing; the other is Six Sigma. They are promoted as different approaches and different thought processes. Yet, upon close inspection, both approaches attack the same enemy and behave like two links within a chain—that is, they are dependent on each other for success. They both battle variation, but from two different points of view. Lean and Six Sigma integration takes two powerful problem-solving techniques and bundles them into a powerful package. The two approaches should be viewed as complements to each other rather than as equivalents of or replacements for each other (Pyzdek, 2000).

In theory, lean manufacturing (hereinafter referred to as "lean") is a winning strategy. In practice, manufacturers that have widely adopted lean practices record performance metrics superior to those achieved by plants that have not adopted lean practices. Those practices cited as lean in a recent industrial survey (Jusko,

1999) include (1) quick changeover techniques to reduce setup time; (2) adoption of manufacturing cells in which equipment and workstations are arranged sequentially to facilitate small-lot, continuous-flow production; (3) just-in-time (JIT) continuous-flow production techniques to reduce lot sizes, setup time, and cycle time; and (4) JIT supplier delivery in which parts and materials are delivered to the shop floor on a frequent, as-needed basis.

Lean evaluates the entire operation of a factory and restructures the manufacturing method to reduce wasteful activities like waiting, transportation, material hand-offs, inventory, and overproduction. It co-locates the processes in sequential order and, in so doing, reduces variation associated with manufacturing routings, material handling, storage, lack of communication, batch production, and so forth. Six Sigma tools, on the other hand, commonly focus on specific part numbers and processes to reduce variation. The combination of the two approaches represents a formidable opponent to variation in that it includes both relayout of the factory and a focus on specific part numbers and processes. The next section discusses the similarities and differences between the two approaches.

Lean Manufacturing and Noise

Lean tackles the most common form of process noise by aligning the organization in such a way that it can begin working as a coherent whole instead of as separate units. Lean manufacturing seeks to co-locate, in sequential order, all the processes required to produce a product. Instead of focusing on the part number, lean methodology focuses on product flow and on the operator. How can the product be produced in the least amount of time, and how can there be standardization of operator methods along with transfer to the operator of what he or she needs? Flow-focused cells in an organization reduce the communication barriers that exist at the numerous interfaces between operations and greatly reduce the time to achieve a completed part.

Figure 1.2 illustrates Denecke's view that a lean approach attacks variation differently than a Six Sigma system does (Denecke, 1998). For example, lean manufacturing focuses on variation associated with different lean practices employed to speed production and improve quality. Examples of individual practices that can contribute to noise include maintenance, cleanliness, process sequence/co-location issues, and so on. Lean drives two key components of quality: process speed and the feedback of information. If all required machinery is placed together and parts are flowed one piece at a time, any quality problem should be identified by downstream processes immediately. Furthermore, the co-location of multidisciplinary operators allows for speedy communication and problem solving without outside intervention. With cross-training, operators understand how upstream processes affect downstream quality.

Lean manufacturing changes the set of problems the organization must address and increases the price of failure. The organization is forced to respond to the lean definition of waste in the process—travel distance, wait time, non-value-added cycle time, setup time, inventory levels, overproduction, and so on. Setup time is an especially important measure in the lean manufacturing line. Once the first machine starts setting up for a different part configuration, the rest of the line becomes starved for parts—in other words, setup time becomes cumulative. The time involved in setting up the entire cell is the sum of all the individual setup times. A two-hour setup per machine may result in 20 hours of downtime for the entire cell. A host of problems and opportunities that the organization could afford to hide in the past now become critical owing to this cumulative effect. A Six Sigma organization needs to understand the special demands of lean manufacturing if it plans to move toward a quality system that embraces both approaches.

Six Sigma and Noise

The data-driven nature of Six Sigma problem solving lends itself well to lean standardization and the physical rearrangement of the factory. Lean provides a solid foundation for Six Sigma problem solving where the system is measured by deviation from and improvements to the standard.

One of the most difficult aspects of Six Sigma work is implementing and sustaining gains. Even experts in the application of Six Sigma techniques struggle to effectively safeguard the production process from noise. Typically, the employee responsible for implementing Six Sigma is such a strong agent of change that when he or she leaves the area in question to work on a new problem, a vacuum of sorts is created. The process variation may build up over time and allow quality problems to resurface. As Figure 1.2 illustrates, the types of noise that resurface typically involve part variation and overall process variation—for example, poor yield, scrap and rework, raw material variability from multiple vendors, and so forth. Lean provides an excellent safeguard for Six Sigma practitioners against many common sources of noise.

Synergy

Lean and Six Sigma, working together, represent a formidable weapon in the fight against process variation. Six Sigma methodology uses problem-solving techniques to determine how systems and processes operate and how to reduce variation in processes. In a system that combines the two philosophies, lean creates the standard and Six Sigma investigates and resolves any variation from the standard.

With the lean emphasis on standardization, Six Sigma practitioners can be 10 times more effective at tackling process noise. If a number of traditional

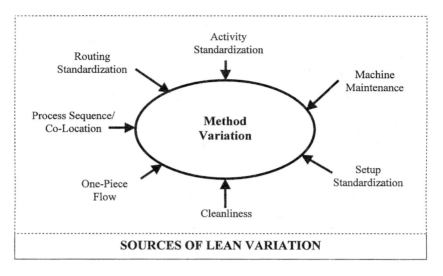

SOURCES OF LEAN VARIATION

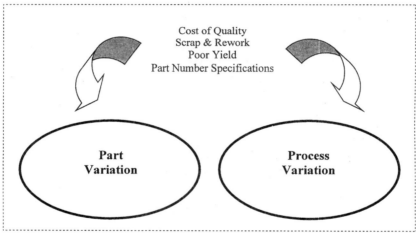

Figure 1.2 Variation as viewed by lean manufacturing and Six Sigma [Adapted from Denecke (1998).]

noise factors have been eliminated by the lean activity, the rest are now easily identifiable through simple observation of the cell. The new system enables rapid problem solving since the Six Sigma practitioner does not have to chase parts through the factory. The hidden factory, now exposed, shows a number of new opportunities for variation—for example, cycle time variation, queuing of parts, downtime of the cell, unpredictable walk patterns, special skill requirements, nonstandard operations, quality problems, and so on. The Six Sigma practitioner can now focus quickly on the problem, assemble the natural work team, collect data in one contained area, compare the results to the standard, and begin the improvement project. The Six Sigma practitioner becomes proactive, continuously seeking to improve upon the standard by expanding on the current process knowledge. It is the responsibility of the organization to go lean so that those implementing Six Sigma are used not as "firefighters" but as proactive and productive problem solvers.

Summary

Again, in our opinion the differences between how Six Sigma and lean manufacturing define and address a project are even less pronounced than what is suggested by the article upon which the preceding discussion is based. We believe that Six Sigma within any organization should have metrics, such as cycle time, that lead to the application of lean manufacturing when it is appropriate. In addition, the techniques of Six Sigma should be applied within an organization's processes to reduce defects, which can be a very important prerequisite to the success of a lean manufacturing project.

1.6   FREQUENTLY ASKED QUESTIONS ABOUT SIX SIGMA

What Is Six Sigma?

Six Sigma provides companies with a series of interventions and statistical tools that can lead to breakthrough profitability and quantum gains in quality, whether the products of a company are durable goods or services. Sigma, σ, is a letter in the Greek alphabet used to denote the standard deviation of a process. Standard deviation measures the variation or amount of spread about the process average. We would like to obtain outputs from processes such that the overall variability of the process is well within specification limits. Sigma quality level, discussed later in this book, is sometimes used to describe the output of a process. A higher sigma quality level value is better. A Six Sigma quality level is said to equate to a 3.4 parts per million outside specification limits. Minitab (2000) claims that today most organizations operate between two and three sigma. Blakeslee (1999) claims that U.S. manufacturing firms frequently attain four sigma quality levels, whereas service firms often operate at quality levels of one or two sigma.

What Is the Practical Difference Between Three and Four Sigma Quality Levels?

To a first approximation, a three sigma quality level is equivalent to 1.5 misspelled words per page in a book such as this one. A six sigma quality level is equivalent to one misspelled word in all of the books in a small library. The differences are huge. The difference stems from the nonlinear relationship between these defect rates and sigma quality levels.

Table 1.6 (Texas Instruments, 1992) provides another basis for comparison among sigma quality levels.

TABLE 1.6   What Is the Magnitude of Difference Between Sigma Levels?

Sigma Level	Area	Spelling	Time	Distance
1	Floor space of the the Astrodome	170 misspelled words per page in a book	31¾ years per century	From here to moon
2	Floor space of a large supermarket	25 misspelled words per page in a book	4½ years per century	1½ times around the world
3	Floor space of a small hardware store	1.5 misspelled words per page in a book	3½ months per century	Coast-to-coast trip
4	Floor space of a typical living room	1 misspelled word per 30 pages (typical book chapter)	2½ days per century	45 minutes of freeway driving
5	Size of the bottom of your telephone	1 misspelled word in a set of encyclopedias	30 minutes per century	1 trip to the local gas station
6	Size of a typical diamond	1 misspelled word in all the books in a small library	6 seconds per century	Four steps in any direction
7	Point of a sewing needle	1 misspelled word in all the books in several large libraries	One eye-blink per century	Inch

Courtesy of Texas Instruments.

What Kind of Return on Investment (ROI) Can I Expect from Six Sigma Training?

Average project savings can vary between $50,000 and $200,000. Six Sigma Practitioners, with 100% of their time allocated to projects, can execute five or six projects during a 12-month period, potentially adding an average of over $1 million to annual profits.

Why Should I Give Six Sigma Serious Consideration?

The customers that form the base of today's world market are sending a clear and undeniable message: produce high-quality products at lower costs with greater responsiveness. To compete in a world market, a company needs to move toward a Six Sigma level of performance. Many of the most profitable and admired companies in the world have set goals within a Six Sigma context, and many moving toward Six Sigma levels of performance have saved billions of dollars. The changes that occur within corporations striving for Six Sigma quality translate to bottom-line benefits.

What Models Are Used to Implement Six Sigma?

There are two basic models for implementing Six Sigma. One model is based on teaching the tools of Six Sigma. In this model little attention, if any, is given to building an organizational infrastructure that supports Six Sigma. Instead, emphasis is typically given to the mechanics of tool execution, as opposed to when and how a tool should be implemented and integrated with other tools.

The other model is project based: the tools are taught and then applied to projects that are defined before sessions begin. In this model, Six Sigma infrastructure issues need to be addressed; however, this effort is sometimes not given the emphasis it deserves. Organizations can achieve more significant bottom-line benefit with this approach, which we will focus on throughout this book.

What Are the Roles Required in a Six Sigma Implementation?

The term "Six Sigma" originated at Motorola, where the Six Sigma methodology was developed and refined. Eventually the approach was adopted and expanded upon by other organizations. Terms were coined at Motorola and elsewhere to describe various roles in the Six Sigma process. To avoid any controversy or conflict over the use of some of these historical terms, a number of organizations have chosen to refer to their Six Sigma activities using different names. For example, AlliedSignal refers to Six Sigma as "operational excellence."

Typically, special names are given to people within an organization who have Six Sigma roles. A few of the names for these roles given by GE, Motorola, and other companies are Process Owner or Sponsor, Champion, Master Black Belt, Black Belt, and Green Belt (Cheek, 1992; General Electric, 1996; General Electric, 1997; Lowe, 1998; Slater, 1999; Pyzdek, 1999; Harry, 2000). These roles are described in more detail later in this book. Note in Table 1.7 the distinction between a "supporter" and a "champion" of Six Sigma (Schiemann and Lingle, 1999). Both roles are important, but they differ in several crucial respects.

TABLE 1.7   The Difference Between a Champion and a Supporter

Champion	Supporter
• Recruits new supporters of Six Sigma	• Maintains support of the Six Sigma process
• Anticipates and addresses Six Sigma problems before they occur	• Reacts to and helps solve Six Sigma problems that occur
• Speaks inspirationally about Six Sigma	• Speaks supportively of Six Sigma
• Actively seeks opportunities to show support of Six Sigma	• Avoids public actions which appear nonsupportive of Six Sigma
• Initiates important Six Sigma activities	• Participates in planned Six Sigma activities

What Is Six Sigma Training Like?

For those who are to implement Six Sigma techniques, we suggest training that involves a total of four months of coursework and on-the-job preparation. Each of the monthly sessions begins with a week of intensive coursework. Class size ranges from 10 to 25, with 20 the target. Coursework involves use of a laptop computer throughout the week, with a computer software package such as Minitab (2000) statistical software and the Microsoft Office Suite installed.

Participants are assigned a project before they begin training, and are required to make brief progress reports on their projects during the second, third, and fourth week of training. They are also required to submit a written report, when the training is complete, outlining what has been accomplished on the project during the four-month period. Participants are also expected to confirm the declared cost savings resulting from the project, in consultation with a company financial officer.

How Much Quality Experience and Statistical Knowledge Is Required of a Six Sigma Instructor?

When selecting people to become instructors, many organizations focus primarily on the degrees and certifications that an individual has achieved. Rather than focusing on a list of credentials, it is perhaps more important to focus on obtaining instructors who will perform well in a classroom environment with high-quality instructional material. Instructors also need to have practical experience that will allow them to bridge the gap between classroom examples and the application of techniques to specific situations—such as those that students present during training and coaching sessions.

What Traits Should I Look for in Full-Time Practitioners of Six Sigma?

When selecting candidates to become full-time practitioners of Six Sigma, look for these characteristics:

- Fire in the belly; an unquenchable drive to improve products and processes
- "Soft skills" and the respect of others within the organization; an ability to create, facilitate, train, and lead teams to success
- Ability to manage projects and reach closure; a persistent drive toward meaningful bottom-line results and timely completion of projects
- Basic analytical skills; an ability to learn, apply, and wisely link Six Sigma tools, some of which may utilize computer application programs

It should be noted that this list does not call for a knowledge of statistical techniques. Our belief is that trainees with the characteristics listed above will easily pick up statistical techniques if provided with good instructional material and training.

Do We Need Full-Time Personnel for Six Sigma?

Implementing a Six Sigma process requires a considerable investment in infrastructure. The number of projects that can be undertaken is a function of the level of infrastructure support allocated to Six Sigma. As with many other programs and processes, there is a minimum critical mass necessary for success. Six Sigma needs to be a "top-down" process, primarily because executives have the mandate to determine what gets done in an organization. Six Sigma projects should be tied strategically to the business plan and should focus on satisfying customers and delivering world-class performance based on multidimensional metrics. An organization that does not devote personnel full-time to implementing Six Sigma risks having its Six Sigma initiative become diluted.

What Kind of "Soft" Skills Are Required to Be a Successful Six Sigma Practitioner?

Soft skills are as important as, if not more important than, hard skills—and many are more difficult to acquire. For reasons that are not easy to understand, many people who are skilled in technical and statistical methods (left-brain dominance) lack the abilities and finesse in human relations and team building (right-brain dominance). Some important soft skills are the following:

- The ability to be an effective facilitator
- The ability to effectively charter Six Sigma teams
- The ability to provide training covering a wide variety of quality tools
- The ability to explain technical details in layman's terms
- The ability to talk candidly with and effectively influence senior management
- The ability to talk candidly with and effectively influence line worker personnel
- A balanced concern for people issues as well as programmatic issues
- Unquestioned integrity
- The ability to mediate strong disagreements among varied constituencies
- The ability to disassociate from bias on how things were done in the past
- The ability to deal with complex situations and complicated personalities diplomatically
- The ability to make effective presentations to executives, managers, and workers

What Infrastructure Issues Will I Face if I Adopt a Six Sigma Approach?

Refer to Chapter 8 in this book for a complete answer to this important question.

2

SIX SIGMA BACKGROUND AND FUNDAMENTALS

Six Sigma was first espoused by Motorola in the 1980s. Recent Six Sigma success stories, primarily from the likes of General Electric, Sony, AlliedSignal, and Motorola, have captured the attention of Wall Street and have propagated the use of this business strategy. A Six Sigma initiative is designed to change the culture in a company through breakthrough improvements by focusing on out-of-the-box thinking in order to achieve aggressive, stretch goals. The Six Sigma initiative has typically contributed an average of six figures per project to the bottom line.

In this chapter we discuss the experiences of other organizations with the initiative and also discuss a statistical definition of Six Sigma.

2.1 MOTOROLA: THE BIRTHPLACE OF SIX SIGMA

Bill Wiggenhorn is senior vice president of Motorola Training and Education, and president of the distinguished Motorola University. He provided a fore-word for the newly released book on Six Sigma, *Implementing Six Sigma: Smarter Solutions using Statistical Methods* (Breyfogle, 1999). We quote from his comments:

> In preparing these remarks, I reflected on the beginnings of Six Sigma and the pioneers who had the courage, intellect, and vision to make it a reality. The father of Six Sigma was the late Bill Smith, a senior engineer and scientist in our com-

31

munications products business. It was Bill who crafted the original statistics and formulas that were the beginnings of the Six Sigma culture. He took his idea and his passion for it to our CEO at the time, Bob Galvin. Bob could see the strength of the concept, but reacted mostly in response to Bill's strong belief and passion for the idea. He urged Bill to go forth and do whatever was needed to be done to make Six Sigma the number one component in Motorola's culture. Not long afterwards, Senior Vice President Jack Germaine was named as quality director and charged with implementing Six Sigma throughout the corporation. Jack had a big job and was armed with few resources. So he turned to Motorola University to spread the Six Sigma word throughout the company and around the world. Soon, Six Sigma training was required for every employee. The language of quality became the common Motorola language. Whether it was French, Arabic, or Texan, everyone understood the six steps, defect measurement and elimination, and parts per million. The training and concepts were not limited to the factory floor. Every single person was expected to understand the process and apply it to everything that they did. Some shied away, saying they were exempt from manufacturing processes. Some challenged the soundness of the statistics. Even so, the company stood behind the commitment and the mandate.

The result was a culture of quality that permeated throughout Motorola and led to a period of unprecedented growth and sales. The crowning achievement was being recognized with the Malcolm Baldrige National Quality Award. At the time, Bill remarked that Motorola was the only company who could receive the award. We were, to his knowledge, the only applicant who had the processes, measurements, and documentation in place to tell our quality story.

Motorola launched its "Six Sigma Quality Program" on January 15, 1987. The program was kicked off with a speech by Motorola's chief executive officer, Bob Galvin, that was distributed in the form of both a letter and a videotape. The company set a five-year target to achieve Six Sigma. By March 1988, Motorola University had begun offering a course on implementing Six Sigma that was aimed primarily at services rather than products. Within a few months of the initial training, teams were starting improvement projects to reach their new corporate quality goals.

2.2   GENERAL ELECTRIC'S EXPERIENCES WITH SIX SIGMA

General Electric (GE) has the unique distinction of being at the head of the list of Fortune 500 companies based on market capitalization. What does market capitalization mean? It means that if someone multiplies GE's outstanding shares of stock by its current market price per share, GE is the highest-valued company listed on any of the numerous U.S. stock exchanges. This monetary value exceeds the gross domestic product of many nations in the world.

General Electric CEO Jack Welch describes Six Sigma as "the most challenging and potentially rewarding initiative we have ever undertaken at General Electric" (Lowe, 1998). The GE 1997 annual report states that Six Sigma delivered more than $300 million to its operating income. In 1998, this number increased to $750 million. GE listed the following examples as typical Six Sigma benefits (General Electric,1997):

- "Medical Systems described how Six Sigma designs have produced a 10-fold increase in the life of CT scanner x-ray tubes—increasing the 'uptime' of these machines and the profitability and level of patient care given by hospitals and other health care providers."

- "Super-abrasives—our industrial diamond business—described how Six Sigma quadrupled its return on investment and, by improving yields, is giving it a full decade's worth of capacity despite growing volume—without spending a nickel on plant and equipment capacity."

- "Our railcar leasing business described a 62% reduction in turnaround time at its repair shops: an enormous productivity gain for our railroad and shipper customers and for a business that's now two or three times faster than its nearest rival because of Six Sigma improvements. In the next phase across the entire shop network, black belts and green belts, working with their teams, redesigned the overhaul process, resulting in a 50% further reduction in cycle time."

- "The plastics business, through rigorous Six Sigma process work, added 300 million pounds of new capacity (equivalent to a 'free plant'), saved $400 million in investment, and will save another $400 million by 2000."

Six Sigma training has permeated GE, and experience with Six Sigma implementation is now a prerequisite for promotion to all professional and managerial positions. Executive compensation is determined to a large degree by one's proven Six Sigma commitment and success. GE now boasts slightly under 4,000 full-time, trained Black Belts and Master Black Belts. They also claim more than 60,000 part-time Green Belts who have completed at least one Six Sigma project (Pyzdek, 1999).

2.3   EASTMAN KODAK'S EXPERIENCE WITH SIX SIGMA

The following is reprinted with permission from Eastman Kodak Company:

"Eastman Kodak Company has been actively involved in the global quality transformation since the late 1950s (Gabel and Callanan, 1999). In the early 1980s they introduced their Quality Improvement Facilitator (QIF) Program. In the early

1990s they joined a Six Sigma consortium involving Motorola, Texas Instruments, Digital Equipment Corporation, and IBM. Since that time, they have been actively pursuing a Six Sigma process performance throughout the company.

"The Office of Six Sigma Programs is part of the Corporate Quality Organization of Eastman Kodak Company. This office helps define and refine the portfolio of quality training programs and functions within the company. In 1999, Kodak had three corporate training and certification programs for building Six Sigma capabilities within the company: the redesigned Quality Improvement Facilitator Program, the Black Belt Program, and the Management Black Belt Program.

"The QIF Program is designed to develop general practitioners in quality that can support all organizational improvement efforts. Lead operators and 1st level supervisors engage in seven weeks of training in the areas of interpersonal skills, leadership skills, basic quality and statistics, and other technical topics. Project work is required as part of the certification process. These generalists are then qualified to coach, teach, and facilitate as active members of any organizational improvement teams.

"The Black Belt Program is designed to develop quality specialists with advanced training and skill in statistics. Their role is to lead major breakthrough improvement efforts. Candidates for this program are drawn from the ranks of experienced technical leaders within Kodak. The five weeks of training begin with a week focused on interpersonal skills, followed by four weeks of training based on the traditional Six Sigma MAIC model of measure, analyze, improve and control. One week is devoted to each of these four MAIC areas. The curriculum provides extensive coverage in the areas of interpersonal skills, advanced quality and statistics, and other technical topics. A certification project is also required.

"The Management Black Belt Program is designed to enable 2nd level supervisors up to and including upper management to not only talk-the-talk of Six Sigma, but also more importantly to walk-the-walk. Training involves approximately one week in the classroom covering the general topics of basic quality and statistics. A certification project is required.

"To date, Kodak has saved over $100 million on the certification projects associated with these three programs. The Six Sigma training initiative started at their Rochester-based operations, but has since been expanded worldwide. The Black Belts, Management Black Belts, and Quality Improvement Facilitators are highly valued resources within the Kodak community. Most of the original Six Sigma activity to date has been in the manufacturing area, but has recently expanded into business and transactional processes as well. The R&D Division is currently applying Six Sigma philosophy and methods to the organization's internal research and development activity.

"Kodak reports that five factors have been critical to the success of their Six Sigma training initiative:

- Management support
- Quality of the work environment

- Quality of the Six Sigma and QIF candidates
- Consistency across quality programs
- Effectiveness of the program instructors"

2.4 ADDITIONAL EXPERIENCES WITH SIX SIGMA

A *USA Today* article presented differences of opinion about the value of Six Sigma in "Firms Air for Six Sigma Efficiency" (Jones, 1998). One stated opinion was that Six Sigma is "malarkey," whereas Larry Bossidy, CEO of Allied Signal, countered with "The fact is, there is more reality with this [Six Sigma] than anything that has come down in a long time in business. The more you get involved with it, the more you're convinced." The following are some other quotes from the article:

- "After four weeks of classes over four months, you'll emerge a Six Sigma 'black belt.' And if you're an average black belt, proponents say you'll find ways to save $1 million each year."
- " Six Sigma is expensive to implement. That's why it has been a large-company trend. About 30 companies have embraced Six Sigma including Bombardier, ABB [Asea Brown Boveri] and Lockheed Martin."
- "Nobody gets promoted to an executive position at GE without Six Sigma training. All white-collar professionals must have started training by January. GE says it will mean $10 billion to $15 billion in increased annual revenue and cost savings by 2000 when Welch retires."
- "Raytheon figures it spends 25% of each sales dollar fixing problems when it operates at four sigma, a lower level of efficiency. But if it raises its quality and efficiency to Six Sigma, it would reduce spending on fixes to 1%."
- "It will keep the company [AlliedSignal] from having to build an $85 million plant to fill increasing demand for caperolactan used to make nylon, a total savings of $30–$50 million a year."
- "Lockheed Martin used to spend an average of 200 work-hours trying to get a part that covers the landing gear to fit. For years employees had brainstorming sessions, which resulted in seemingly logical solutions. None worked. The statistical discipline of Six Sigma discovered a part that deviated by one-thousandth of an inch. Now corrected, the company saves $14,000 a jet."
- "Lockheed Martin took a stab at Six Sigma in the early 1990s, but the attempt so foundered that it now calls its trainees 'program managers' instead of black belts to prevent in-house jokes of skepticism ... Six Sigma

is a success this time around. The company has saved $64 million with its first 40 projects."

- "John Akers promised to turn IBM around with Six Sigma, but the attempt was quickly abandoned when Akers was ousted as CEO in 1993."

- "Marketing will always use the number that makes the company look best ... Promises are made to potential customers around capability statistics that are not anchored in reality."

- "Because managers' bonuses are tied to Six Sigma savings, it causes them to fabricate results and savings turn out to be phantom."

- "Six Sigma will eventually go the way of other fads, but probably not until Welch and Bossidy retire."

As apparent from the vast differences in opinion listed, Six Sigma can be a great success or failure depending upon how it is implemented.

2.5 RECENT ADVANCES

Recent Six Sigma success stories, primarily from the likes of General Electric, Sony, AlliedSignal, and Motorola, have captured the attention of Wall Street and have propagated the use of this business strategy. The Six Sigma strategy involves the use of statistical tools within a structured methodology for gaining the knowledge needed to achieve better, faster, and lower-cost products and services. The repeated, disciplined application of the master strategy on project after project, where the projects are selected based on key business issues, is what drives money to the bottom line, resulting in increased profit margins and impressive return on investment from the Six Sigma training. The Juran Institute (Godfrey, 1999) emphasizes the point that an organization's vision and strategy must be the compass directing Six Sigma project selection and measurement. The typical Six Sigma initiative has contributed an average of six figures per project to the bottom line when properly implemented.

Six Sigma project specialists are sometimes called "Black Belts," "top guns," "change agents," or "trailblazers," depending on the company deploying the strategy. These people are trained in the Six Sigma philosophy and methodology and are expected to accomplish at least four projects annually, which should deliver at least $500,000 annually to the bottom line. A Six Sigma initiative in a company is designed to change the culture through breakthrough improvements by focusing on out-of-the-box thinking in order to achieve aggressive, stretch goals. Ultimately, Six Sigma, if deployed properly, will infuse intellectual capital into a company and produce unprecedented knowledge gains that translate directly into bottom-line results (Kiemele, 1998).

We believe that many organizations will find that Six Sigma implementation is the best thing that ever happened to them. Unfortunately, we predict that some companies embarking upon the Six Sigma path will find the journey difficult. Success depends on how Six Sigma is implemented. The road map in this book can lead an organization away from a Six Sigma strategy built around "playing games with the numbers" to a strategy that produces long-lasting process improvements yielding significant bottom-line results.

2.6   THE STATISTICAL DEFINITION OF SIX SIGMA

The statistics associated with Six Sigma may initially seem difficult, but they are relatively simple to grasp and apply, if a little effort is given to understanding them. The reader is referred to the glossary at the end of this book for a definition of any new terms. To define Six Sigma statistically, we will work with two concepts, specification limits and the normal distribution.

Specification Limits

Specification limits are the tolerances or performance ranges that customers demand of the products or processes they are purchasing. An example of a specification might be the size of a given circular hole drilled into a circuit board in a manufacturing plant. For purposes of illustration, we will refer to this hole as "XYZ." A circuit board may have hundreds of holes that have specified diameters. If hole XYZ is too large, it will tend to accumulate excess solder that could lead to other problems. If it is too small, the "metal lead" that gets inserted into the hole and subsequently soldered into place may be too big to fit properly. The customer desires an XYZ hole diameter of exactly xyz inches, but will accept XYZ holes with diameters that fall within the range between the customer-specified lower specification limit [LSL] and upper specification limit [USL]. Why? Because variability is so ubiquitous in the real world that we have to allow specification limits that permit some degree of imprecision in our ability to drill xyz-inch-diameter holes.

Figure 2.1 illustrates specification limits as the two major vertical lines in the figure. The target value, although not shown in the figure, is typically at the exact center between the USL and LSL. These specification limits are totally independent of the bell-shaped curve of the normal distribution, also shown in the figure. The important thing to realize about the range between the USL and the LSL is that the customer expects every XYZ hole drilled in their circuit board to have a diameter that falls somewhere within that range and, hopefully, falls exactly in the center of that range. Is it acceptable for an occasional XYZ hole to have a diameter equal to one of the specification limits? It depends on the customer, who will decide whether or not the extreme values at the specification limits are acceptable quality levels.

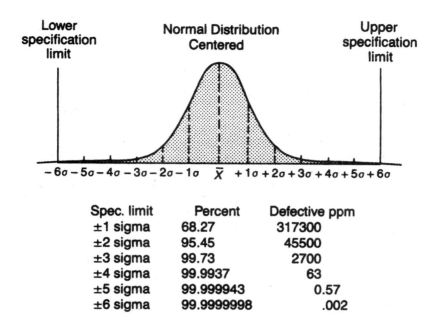

Spec. limit	Percent	Defective ppm
±1 sigma	68.27	317300
±2 sigma	95.45	45500
±3 sigma	99.73	2700
±4 sigma	99.9937	63
±5 sigma	99.999943	0.57
±6 sigma	99.9999998	.002

Figure 2.1 With a central normal distribution between Six Sigma limits, only two devices per billion fail to meet the specification target. (Copyright of Motorola, used with permission.)

The Normal Distribution

The bell-shaped curve in Figure 2.1 is called the normal distribution, also known as a Gaussian curve. It has a number of properties that make it an extremely useful and valuable tool in both the statistical and quality worlds. The curve is symmetrically shaped and extends from + to – infinity on the X-axis. This normal curve is totally independent of the LSL and USL described above for hole XYZ. The shape of this normal curve depends solely on the process, equipment, personnel, and so on, which can affect the drilling of XYZ holes. This normal curve represents the spread of XYZ hole diameters resulting from the drilling or punching in a circuit board using current equipment, materials, workers, and so forth. The normal curve says nothing about the range of XYZ hole diameters acceptable to the customer. This curve summarizes the empirical quantification for the variability that exists within the XYZ hole manufacturing process.

The dashed vertical lines on the curve in Figure 2.1 represent the number of standard deviation units (σ) a given hole diameter might be from the mean, which is shown as $\bar{x}$ on the x-axis. In the example we are using here, the standard deviation (σ) would be expressed in units of inches. The basic formula for calculating σ is straightforward; however, Breyfogle (1999), in *Implementing Six Sigma* (chapter 11), describes confusion and differences of opinion relative to the calculation and use of this metric within Six Sigma. The tabular information beneath Figure 2.1 indicates the percentage of the area under the normal curve that can be found within ± 1 σ units, ± 2 σ units ... ± 6 σ units centered about the mean, where σ is the Greek symbol for true population standard deviation.

The total area under the normal curve is assumed to be equal to a proportion (decimal value) of one. In layman's terms, this one means that all (100%) possible XYZ holes that are drilled have diameters that range from − infinity to + infinity on the X-axis. Although this normal distribution theoretically allows for the production of an XYZ hole that is 12 feet in diameter, this is a practical impossibility and has such a low probability of occurring that it is irrelevant.

So that we do not have to deal with the infinite properties of the normal distribution, the normal distribution in science, engineering, and manufacturing is typically used so that we only concern ourselves with XYZ hole diameters that fall outside the range of ± 3 standard deviations about the target mean. The holes outside of the customer specification limits are defined as defects, failures, or nonconformities. If we assume for this example that our estimate of σ for XYZ hole diameters is abc inches, then ± 3 σ units of standard deviation is equivalent to ± 3 times abc inches. It is important to remember that xyz inches represents the customer's defined target hole diameter. Our empirical estimate of σ for our XYZ hole–producing process is abc inches.

The tabular data beneath Figure 2.1 indicate that ± 3 σ units of standard deviation represent 99.73 percent of the total area under the normal distribution. The difference of 0.27% (i.e., 100% - 99.73%) is the probability of our plant producing an XYZ hole whose diameter is outside of the ± 3 σ units of standard deviation. If a process is centered, for every 100 XYZ holes drilled, 99.73 of them, approximately 98 holes, will have xyz diameters that fall within ± 3 σ, where σ is equal to abc inches. Figure 2.1 notes that this corresponds to 2,700 defective XYZ holes per million XYZ holes drilled for ± 3σ. For ± 6 σ, the ppm value is 0.002.

Sigma Quality Level

The scenario described above considers the situation where a process is centered. To address "typical" shifts of a process mean from a specification-centered value, Motorola added a shift value ± 1.5 σ to the mean. This shift of the mean is used when computing a process "sigma level" or "sigma quality level," as shown in Figure 2.2. From this figure we note, for example, that a 3.4 ppm rate corresponds to a 6 σ quality level. Figure 2.3 illustrates how sigma quality levels would equate to other defect rates and organizational performances. Figure 2.4 illustrates the impact of the ± 1.5 σ shift.

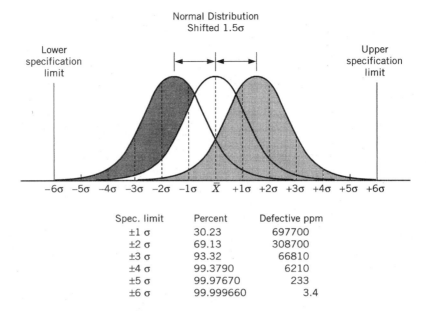

Spec. limit	Percent	Defective ppm
±1 σ	30.23	697700
±2 σ	69.13	308700
±3 σ	93.32	66810
±4 σ	99.3790	6210
±5 σ	99.97670	233
±6 σ	99.999660	3.4

Figure 2.2 Effects of a 1.5 σ shift where only 3.4 ppm fail to meet specifications (Copyright of Motorola, used with permission.)

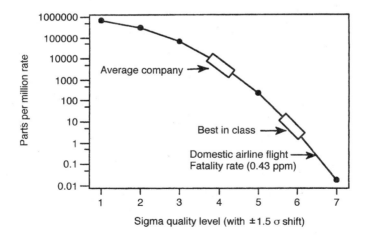

Figure 2.3 Implication of sigma quality level. Part per million (ppm) rate for part or process step, considers a 1.5 σ shift of the mean where only 3.4 ppm fail to meet specification at a six sigma quality level.

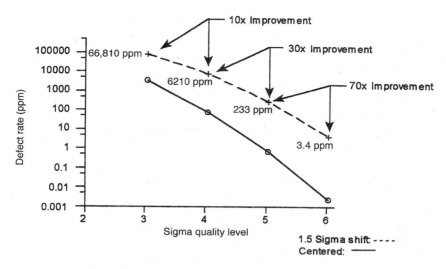

Figure 2.4   Defect rates (ppm) versus sigma quality level.

One point that should be emphasized is that sigma quality levels can be deceptive since they bear an inverse, nonlinear relationship to defect rates. Higher sigma quality levels mean fewer defects per million opportunities, but the relationship is not linear. Figure 2.4 illustrates that an improvement from a three to a four sigma quality level is not the same as an improvement from a five to a six sigma quality level. The first "unit" shift noted above corresponds to a 10x improvement in defect rate, and the latter "unit" shift to a 70x improvement, where these comparisons are based on a ±1.5 σ shifted process. A unit change in sigma quality level does not correspond to a unit change in process improvement as measured by defect rates. A shift in sigma quality level from five to six sigma is a much more difficult improvement effort than a shift in sigma quality level from three to four sigma.

The sigma quality level metric is very controversial. An organization needs to understand the concept behind this metric even if it does not use the metric as a driving force within the organization. We suggest that organizations select other Six Sigma metrics over the Sigma Quality Level metric to monitor improvement activities within their organizations. However, if the need arises to convert between defect rates and sigma quality level, Table 2.1 makes the conversion simple.

Additional Six Sigma metrics are described in Chapter 5.

TABLE 2.1   Conversion Between ppm and Sigma Quality Level

+ / - Sigma Level at Spec. Limit*	Percent within spec.: Centered Distribution	Defective ppm: Centered Distribution	Percent within spec.: 1.5 Sigma Shifted Distribution	Defective ppm: 1.5 Sigma Shifted Distribution
1	68.2689480	317310.520	30.232785	697672.15
1.1	72.8667797	271332.203	33.991708	660082.92
1.2	76.9860537	230139.463	37.862162	621378.38
1.3	80.6398901	193601.099	41.818512	581814.88
1.4	83.8486577	161513.423	45.830622	541693.78
1.5	86.6385542	133614.458	49.865003	501349.97
1.6	89.0401421	109598.579	53.886022	461139.78
1.7	91.0869136	89130.864	57.857249	421427.51
1.8	92.8139469	71860.531	61.742787	382572.13
1.9	94.2567014	57432.986	65.508472	344915.28
2	95.4499876	45500.124	69.122979	308770.21
2.1	96.4271285	35728.715	72.558779	274412.21
2.2	97.2193202	27806.798	75.792859	242071.41
2.3	97.8551838	21448.162	78.807229	211927.71
2.4	98.3604942	16395.058	81.589179	184108.21
2.5	98.7580640	12419.360	84.131305	158686.95
2.6	99.0677556	9322.444	86.431323	135686.77
2.7	99.3065954	6934.046	88.491691	115083.09
2.8	99.4889619	5110.381	90.319090	96809.10
2.9	99.6268240	3731.760	91.923787	80762.13
3	99.7300066	2699.934	93.318937	66810.63
3.1	99.8064658	1935.342	94.519860	54801.40
3.2	99.8625596	1374.404	95.543327	44566.73
3.3	99.9033035	966.965	96.406894	35931.06
3.4	99.9326038	673.962	97.128303	28716.97
3.5	99.9534653	465.347	97.724965	22750.35
3.6	99.9681709	318.291	98.213547	17864.53
3.7	99.9784340	215.660	98.609650	13903.50
3.8	99.9855255	144.745	98.927586	10724.14
3.9	99.9903769	96.231	99.180244	8197.56
4	99.9936628	63.372	99.379030	6209.70
4.1	99.9958663	41.337	99.533877	4661.23
4.2	99.9973292	26.708	99.653297	3467.03
4.3	99.9982908	17.092	99.744481	2555.19
4.4	99.9989166	10.834	99.813412	1865.88
4.5	99.9993198	6.802	99.865003	1349.97
4.6	99.9995771	4.229	99.903233	967.67
4.7	99.9997395	2.605	99.931280	687.20
4.8	99.9998411	1.589	99.951652	483.48

4.9	99.9999040	0.960	99.966302	336.98
5	99.9999426	0.574	99.976733	232.67
5.1	99.9999660	0.340	99.984085	159.15
5.2	99.9999800	0.200	99.989217	107.83
5.3	99.9999884	0.116	99.992763	72.37
5.4	99.9999933	0.067	99.995188	48.12
5.5	99.9999962	0.038	99.996831	31.69
5.6	99.9999979	0.21	99.997933	20.67
5.7	99.9999988	0.012	99.998665	13.35
5.8	99.9999993	0.007	99.999145	8.55
5.9	99.9999996	0.004	99.999458	5.42
6	99.9999998	0.002	99.999660	3.40

*Sometimes referred to as sigma level or sigma quality level when considering process shift.

Reproduced from Breyfogle (1999), with permission

3

SIX SIGMA NEEDS ASSESSMENT

An organization might have a set of balanced scorecard metrics already in place. If this organization were to plot the change in the metrics over time, they typically would find quantitative variability, where, overall, many of the metrics did not improve. The reason for this is that many organizations do not have a structured approach to improve processes that impact key metrics. Six Sigma can help organizations create an infrastructure whereby improvements are made through projects that follow a structured process improvement/reengineering road map that is linked to metrics. However, sometimes this awareness is not enough when "selling" Six Sigma. Within some organizations, one might need to focus more on the Six Sigma monetary benefits when making a needs assessment.

In this chapter we will discuss two measures that are not only useful within the costing of potential S^4 projects but also can help determine whether Six Sigma is right for an organization. These measures are the cost of poor quality and the cost of doing nothing. Again, in this chapter we focus on the monetary justification for implementing Six Sigma; but there can be other driving forces, such as customer satisfaction. Still, it is often easier to get the attention of executive management when discussing monetary implications. A powerful approach is to combine the two, translating customer satisfaction and other strategic metrics into monetary implications for the review of executive management.

Focusing on monetary benefits alone can cause an organization to neglect key situations that could benefit tremendously from an S^4 improvement project.

Sometimes situations exist, such as in governmental agencies or contractors, where ROI is not a primary driver. For these situations we suggest addressing Six Sigma applicability to an organization through the answering of a few questions, which are listed toward the end of this chapter.

3.1   IMPLICATIONS OF QUALITY LEVELS

An organization that is considering Six Sigma might also consider the implications of doing nothing. If any one of the current competitors of an organization, or even a new competitor, achieves six sigma quality levels in the industry or market segment of the organization, the organization's profitable days could be numbered. When Six Sigma is implemented *wisely,* it is not just another quality program. Six Sigma can bring orders of magnitude improvement over historical quality levels and can totally change the rules by which an organization operates. Once someone in an industry implements a successful and enduring Six Sigma business strategy, the way to become or remain the industry leader is to implement Six Sigma smarter than the competition. The important question may not be if, but when, an organization should adopt a Six Sigma business strategy.

Three sigma quality levels that have been prevalent for the past half-century are no longer acceptable. Consider, for example, that a 99.9% yield would result in the following levels of performance in service industries (Schmidt and Finnegan, 1992; Wortman, 1999):

- 20,000 wrong drug prescriptions per year
- One hour of unsafe drinking water monthly
- 22,000 bank checks deducted incorrectly on an hourly basis
- No electricity, water, or heat for 8.6 hours per year
- Two short or long landings at major airports each day
- Two plane crashes per day at O'Hare International Airport
- 500 incorrect surgical procedures per week
- 2,000 lost pieces of mail per hour
- 32,000 missed heartbeats per person per year

To emphasize the implications of a 0.1% defect rate (99.9% yield), consider the following: your heart skips a few hundred beats; your checking account was accidentally debited $1,000 for an expense you did not incur; your vacation flight crashed at O'Hare Airport; your left kidney was removed when it was the right kidney that was cancerous; your Social Security check was eaten by the mail-processing equipment; your glass of water was contaminated with *Salmonella* bacteria. These types of failures, although seemingly small, are significant, and occasionally fatal, to those people who experience them.

If it is a foregone conclusion that these problems need to be prevented before they happen, it follows that there should be a movement toward six sigma quality levels in all *important* products and processes. Granted, some things are not worth the effort to take them to a six sigma quality level. We don't care if 1 out of 100 wooden matches (99.0% quality) breaks when we try to light the patio barbecue. We do care if the flow valve on the barbecue propane tank leaks and causes an explosion when the match does ignite.

3.2 COST OF POOR QUALITY (COPQ)

Today's global financial markets promote reward through short-term success as measured by quarterly net profit and return on investment. Organizations, both public and private, that can virtually eliminate the cost of poor quality can become the leaders of the future. Conway (1992) claims that in most organizations 40% of the total effort, both human and mechanical, is waste. If that waste can be eliminated or significantly reduced, the per-unit price that must be charged for goods and services to yield a good return on investment is greatly reduced, and often ends up being a price that is competitive on a global basis (Cupello, 1999). Hence, one of the characteristics of world-class performers seems to be negligible investment in COPQ. More than one organization has already realized the advantages of using a Six Sigma approach in improving profitability and customer satisfaction.

Federal and state governments require most, if not all, organizations to maintain financial reports and documents that serve as the basis for determining levels of taxation. Even if there were no taxes, organizations would keep accurate financial records to ensure their own long-term and short-term viability. One very important financial record is the variance report, which compares actual expenditures and sales with projected or budgeted expenditures and sales. Significant variances are immediately addressed and corrective action taken so as not to run out of money or product/services. Until the middle of this century few, if any, companies focused their attention on the costs associated with quality items on the income statement and/or balance sheet. Only the most obvious quality department concerns or needs may have been identified and addressed.

In the 1950s and 1960s, some progressive organizations started to evaluate and report on quality costs. This eventually led to a method for defining, measuring, and reporting quality costs on a regular basis—often weekly, monthly, or quarterly. On an income statement the costs of quality are typically found "hidden" under the categories of overhead costs (indirect costs) and direct costs (direct labor and direct materials). But the initial emphasis on quality costs was not fully accurate, as only the most obvious categories of quality costs were addressed. These are shown as the tip of the iceberg in Figure 3.1.

As the quality movement progressed throughout the latter half of this century, it became obvious that the costs associated with quality could represent as much as 15%–25% of total operating costs (Wortman, 1995) and that many of these costs were not directly captured on the income statement or balance sheet. These truly "hidden" quality costs were those shown below the water line in Figure 3.1.

The addition of technical specialists within the quality department helped define and focus on these hidden quality costs. Calculations of large quality costs caught the attention of the executive level of management. Eventually these costs were referred to as the "cost of poor quality" (COPQ), since they represented unsatisfactory products or practices that, if eliminated, could significantly improve the profitability of an organization. Over a period of decades a number of surprising facts surfaced about COPQ (Juran, 1988).

- Quality-related costs were much higher than financial reports tended to indicate (20%–40% of sales).
- Quality costs were incurred not only in manufacturing but in support areas as well.
- While many of these costs were avoidable, there was no person or organization directly responsible for reducing them.

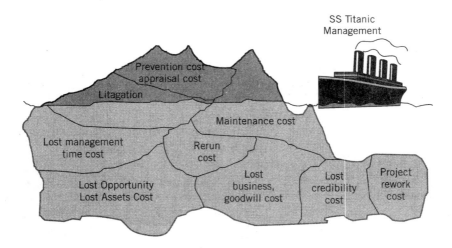

Figure 3.1   Cost of poor quality (Reproduced with permission of RAS Group.)

Eventually, the COPQ was defined in more detail. At the present time, these costs are allocated to three major categories: prevention, appraisal, and failure. In addition, the area of failure cost is typically broken into two subcategories: internal failure and external failure. Prevention costs are devoted to keeping defects from occurring in the first place. Appraisal costs are associated with efforts (such as quality audits) to maintain quality levels by means of formal evaluations of quality systems. Failure costs refer to after-the-fact efforts devoted to products that do not meet specifications or that fail to meet customers' expectations. Table 3.1 gives examples of individual cost elements within each of these major categories (Wortman, 1995).

An S^4 business strategy directly attacks the cost of poor quality (COPQ). Quality cost issues can dramatically affect a business. Wisely applied Six Sigma techniques can help eliminate or reduce many of the issues that affect overall cost. However, management needs to ask the right questions so that these issues are effectively addressed, or set up an infrastructure designed to inherently ask these questions. For example, "beating up" workers and supervisors for high scrap rates is not very effective if the cause of the high rates is inadequate operator training, outdated production equipment, or poor test equipment for measuring quality parameters.

Table 3.1   Examples of Quality Costs

PREVENTION
Training
Capability Studies
Vendor Surveys
Quality Design

APPRAISAL
Inspection and Test
Test Equipment and Maintenance
Inspection and Test Reporting
Other Expense Reviews

INTERNAL FAILURE
Scrap and Rework
Design Changes
Retyping Letters
Late Time Cards
Excess Inventory Cost

EXTERNAL FAILURE
Warranty Costs
Customer Complaint Visits
Field Service Training Costs
Returns and Recalls
Liability Suits

In summary, the concept of COPQ can help identify Six Sigma projects that (1) have the potential for significant monetary savings, (2) are of strategic importance, (3) involve key process output variable (KPOV) issues, and (4) are of vital concern to the customer. It would be ideal if a Pareto chart of the monetary magnitude of the 18 COPQ subcategories listed in Table 3.1 could be created so that areas for improvement could be identified; however, creating such a chart could involve a lot of effort. As a start, one might begin by quantifying the COPQ within one strategic area of a business, perhaps on-site rework, and then later work at quantifying the magnitudes of COPQ within other areas of the business. In a subsequent chapter, we specifically discuss how COPQ ties in to effective project selection.

3.3   COST OF DOING NOTHING

We have addressed the question "Is Six Sigma worth doing?" The flip side of that coin is "What will it cost us if we don't do Six Sigma? What is the cost of doing nothing differently?"

When assessing Six Sigma, leaders should consider the various options: doing nothing, creating a Six Sigma initiative, and creating a Six Sigma business strategy. Let's consider each of these options.

The "doing nothing" option might be the right choice for an organization; however, an organization needs to make this decision after comparing the cost of doing nothing to the cost of doing something.

The "creating a Six Sigma initiative" option typically is viewed by members of the organization as the "program of the month" and risks early abandonment without much benefit. In the worst case, the initiative is dropped after a lot of money and resources are spent on it.

The "Six Sigma business strategy" option has the most benefit if it is executed *wisely*. Our main goal with this book is to give management enough insight into the Six Sigma approach that they can decide whether or not to spend the time and money seriously investigating this approach.

One of the most serious mistakes we can make when trying to answer this question has to do with simple definitions. Many executives confuse efficiency with effectiveness. Efficiency is "doing things right." Effectiveness is "doing the right things." A wise implementation of Six Sigma focuses on "doing the right things right."

3.4   ASSESSMENT QUESTIONS

Often, people want to see a specific example of Six Sigma that closely parallels their own situation so that they can see how they will benefit. However, often there are situations where people have difficulty seeing how Six Sigma techniques are directly applicable to their situation no matter how many examples are shown. This is particularly true of government organizations and government contract organizations, where ROI may not be a primary driving force.

Consider an employee who thinks her organization can benefit from Six Sigma methodology but faces resistance from other employees who do not see the benefits. This person might ask management and other employees questions like those shown in Table 3.2.

Table 3.2 Six Sigma Needs Checklist

	Place a check next to any question that is answered YES!
Do you have multiple "fix-it" projects in a critical process area that seem to have limited or no lasting impact?	
Are you aware of a problem that management or employees are encountering?	
Are you aware of any problem that a customer is having with the products/services your organization offers?	
Do you believe that primary customers might take their business elsewhere?	
Is the quality from competitive products/services better?	
Are your cycle times too long in any process?	
Are your costs too high in any process?	
Do you have concerns that you might be "downsized" from your organization?	
Do you have a persistent problem that you have attempted to fix in the past with limited success?	
Do you have regulatory/compliance problems?	

Any of these questions can be addressed through a project based on Six Sigma thinking. Six Sigma methodology can yield appropriate metrics for these situations and provide guidance on what can be done to resolve them. Through estimating the percentage of revenue lost in any problem areas identified, buy-in can be generated.

PART 2

SIX SIGMA METRICS

4

NUMBERS AND INFORMATION

Organizations often collect data and then react to the data as though they were pieces of information. Consider the following when addressing this behavior (O'Dell and Grayson, 1998):

- Data are simply facts and figures without context or interpretation.
- Information refers to useful or meaningful patterns found in the data.
- Knowledge represents information of sufficient quality and/or quantity that actions can be taken based on the information; actionable information.
- Intellectual capital in its purest form is a subset of knowledge and refers to the commercial value of trademarks, licenses, brand names, patents, etc.

If data are not collected and used wisely, their very existence can lead to activities that are ineffective and possibly even counterproductive. In this chapter we discuss how to create information from the *wise* collection and analysis of data.

4.1 EXAMPLE 4.1: REACTING TO DATA

An organization collects data and reacts whenever an out-of-specification condition occurs. The following shows what can happen when attempts are made to "fix" all out-of-specification problems when they occur within a manufacturing or service environment.

Consider a product that has specification limits of 72–78. An organization might react to collected data this way:

First datum point: 76.2
Everything is OK.
Second datum point: 78.2
Joe, go fix the problem.
Data points: 74.1, 74.1, 75.0, 74.5, 75.0, 75.0
Everything OK—Joe must have done a good job!
Next datum point: 71.8
Mary, fix the problem.
Data points: 76.7, 77.8, 77.1, 75.9, 76.3, 75.9, 77.5, 77.0, 77.6, 77.1, 75.2, 76.9
Everything OK—Mary must have done a good job!
Next datum point: 78.3
Harry, fix the problem.
Next data points: 72.7, 76.3
Everything OK. Harry must have fixed the problem.
Next datum point: 78.5
Harry, seems like there still is a problem.
Next data points: 76.0, 76.8, 73.2
Everything OK—the problem must be fixed now.
Next datum point: 78.8
Sue, please fix the problem that Harry could not fix.
Next data points: 77.6, 75.2, 76.8, 73.8, 75.6, 77.7, 76.9, 76.2, 75.1, 76.6, 76.6, 75.1, 75.4, 73.0, 74.6, 76.1
Everything is great. Give Sue an award!
Next datum point: 79.3
Get Sue out there again. She is the only one who knows how to fix the problem.
Next data points: 75.9, 75.7, 77.9, 78.0
Everything is great again!

A plot of this information is shown in Figure 4.1. From it we can see that the reaction described above does not improve the process or lessen the likelihood of having problems in the future. Figure 4.2 shows a replot of the data as an individuals control chart. The control limits within this figure are calculated from the data, not the customer specifications, as described in Breyfogle (1999), *Implementing Six Sigma* (chapter 10). Since the up-and-down movements are within the upper and lower control limits, it is concluded that there are no

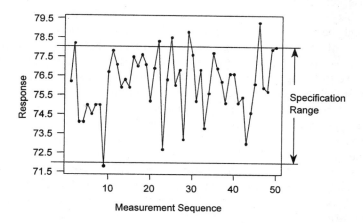

Figure 4.1   Reacting to common cause data as though it were special cause

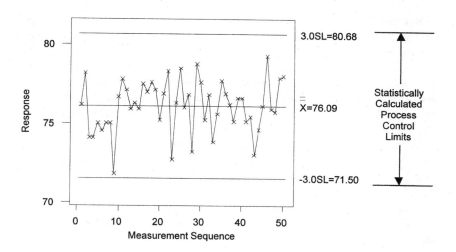

Figure 4.2   Control chart showing variability of data to be from common cause

special causes within the data and that the source of the variability is common cause from the process.

This organization had been reacting to the out-of-specification conditions as though they were special cause. The focus on "fixing" out-of-specification conditions often leads to "firefighting" (an expression that describes the performing of emergency fixes to problems that recur). When firefighting activities involve tweaking the process, additional variability can be introduced, which can further degrade the process rather than improve it.

Both public and private organizations frequently look at human and machine performance data from a perspective similar to those just described and then make judgments based on the data. Production supervisors might constantly review production output by employee, machine, work shift, day of the week, product line, and so forth. In the service sector, an administrative assistant's daily output of letters and memos may be monitored. In call centers around the world, the average time spent per call may be monitored and then used to counsel low-performing employees. The efficiency of computer programmers may be monitored through the tracking of "lines of code produced per day." In a legal department, the number of patents secured on the company's behalf during the last fiscal year may be compared to previous annual rates. Whenever an organization reacts to individual situations that do not meet requirements and specifications, rather than look at the system of "doing work" as a whole, they could be reacting to common cause situations as though they were special cause.

To illustrate this point further, let's look at an organization that monitors the frequency of safety memos. Consider a memo written indicating that the number of "accidents involving injuries" during the month of July was 16, up by 2 from a year ago. The memo declares this increase in accidents to be unacceptable and asks all employees to watch a mandatory, 30-minute safety video by the end of August. At an average wage rate of $10 per hour, the company payroll of 1,500 employees affects the August bottom line by $7,500, which doesn't include wasted time as employees get to and from the conference room. Nor does it take into account the time spent issuing memos reminding people to attend, reviewing attendance rosters looking for laggards, and so forth. How effective was the training video at eliminating the common cause of injury?

We are not saying safety and productivity are not important to an organization. We are saying, as did Dr. Deming, that 94% of the output of a person or machine is a result of the system that management has put in place for use by the workers. If performance is poor, 94% of the time the system must be modified for improvements to occur. Only 6% of the time are problems due to special causes. Knowing the difference between special and common cause variation can affect how organizations react to data and the success they achieve employing a Six Sigma strategy. To reduce the frequency of employees' safety accidents from common cause, an organization needs to look at its systems collectively over a long period to determine what should be done to improve its processes. Reacting to data from an individual month that do not meet a criterion can be both counterproductive and very expensive.

Theodore Levitt (1991) notes that "few things are more important for a manager to do than ask simple questions." One simple question that should be repeated time and time again as Six Sigma implementation proceeds is this: "Is the variation I am observing 'common cause' or 'special cause' variation?" The answer to this question can provide much insight on the actions managers and employees take in response to process and product information. Associatively, those actions have a tremendous impact on employee motivation and self-esteem.

Common cause variability may or may not affect whether a process produces a product or service that meets the needs of customers. We don't know for sure until we compare the process output collectively relative to the specification. This is much different from reacting to individual points that do not meet specification limits. When treating common cause data collectively, emphasis should be given to improving the process. When reaction is made to an individual point that is beyond specification limits for a common cause situation, attention is given to what happened relative to this individual point as though it were a "special" condition, not the process information collectively.

To reiterate, when variation of the "common cause" variety causes out-of-specification conditions, that doesn't mean nothing can or should be done about it. What it does mean is that you need to improve the process, not firefight individual situations that happen to be out of specification. However, it is first essential to identify whether the condition is common or special cause. If the condition is common cause, data are used collectively and the frequency of how the process will perform relative to specification needs is examined. If an individual point is determined to be "special cause" from a process point of view, what was different about this point has to be addressed.

One of the most effective quality tools for distinguishing between common cause and special cause variation is the control chart. Care must be taken with control charts so that managers and workers do not misunderstand and misuse them to the detriment of product and process quality. In the next section, we give advice on how to effectively use control charts in order to reduce the frustration and expense associated with common "firefighting" activities.

4.2 PROCESS CONTROL CHARTING AT THE "30,000-FOOT LEVEL"

Consider the view when looking out an airplane window. When the airplane is at an elevation of 30,000 feet, passengers get a "big picture" view of the landscape. However, when the airplane is at 50 feet, passengers view a much smaller portion of the landscape. Similarly, a "30,000-foot level" control chart gives a macro view of a process, while a "50-foot level" control chart gives more of a micro view of some aspect of the process.

Within training sessions, control charts are most often taught to timely identify special causes within the process at the "50-foot level." An example of this form of control is to timely identify when temperature within a process goes out of control so that the process can be stopped and the temperature-variable problem fixed before a large amount of product is produced with unsatisfactory characteristics.

However, control charts are also very useful at a higher level, where focus is given to directing activities away from firefighting the problems of the day. We suggest using these charts to prevent the attacking of common cause issues as though they were special cause. We want to create a measurement system such that *"fire prevention"* activities are used to address common cause issues where products do not consistently meet specification needs. Our S^4 road map gives a structured approach for "fire prevention" through the *wise* use and integration of Six Sigma tools.

Unlike what is called for by "50-foot level" control charts, we suggest an infrequent sampling to capture how the process is performing overall relative to customer needs. When the sampling frequency is long enough to span all "short-term" process noise inputs, such as raw material differences between days or daily cycle differences, we call this a "30,000-foot level" control chart. At this sampling frequency, we would examine only one sample and plot its response on an individuals control chart.

When first introduced to this concept of infrequent sampling, an individual's typical concern is that this measurement does not give any insight into what should be done to improve the process. The concern is understandable, but the purpose of this type of measurement is not to determine what should be done to improve the process, but rather:

- To determine if "special cause" or "common cause" conditions exist from a "30,000-foot level" vantage point.
- To collect data that provides a "long-term" view of the capability of the process relative to meeting the needs of customers.

We do not suggest that practitioners get bogged down, at the outset, trying to collect a lot of data with the intent of identifying a cause-and-effect relationship that shows the source of their problem. We do suggest that this form of data collection can be appropriate after a problem has been identified and after teams have determined what they think should be monitored.

Process characteristics that are deemed important and require monitoring for the purpose of determining cause-and-effect relationships need to be identified up front during "brainstorming activities" within the S^4 measurement phase. We refer to this collection of various activities—such as process mapping, cause-

and-effect diagram, cause-and-effect matrix, and failure mode and effects analysis (FMEA)—as "wisdom of the organization" outputs.

To summarize, when selecting a rational subgroup to create a "30,000-foot" control chart, the organization needs to create a sampling plan that will give a long-term view of process variability. A sampling plan to create a baseline of a process might be to randomly select one daily KPOV response from a data set compiled during the last year. An individuals control chart could then reveal whether the process has exhibited common cause or special cause conditions. Action to investigate and resolve the root cause of special cause conditions may then be appropriate.

Many special cause conditions appear on a control chart with too frequent sampling. These conditions could be viewed as noise to the system when examined using a less frequent sampling approach. We believe that long-term regular perturbations should be viewed as a common cause of the system/process and should be dealt with accordingly, if they impact KPOV variation. That is, they should be addressed looking at the process inputs/outputs collectively rather than as individual special cause conditions.

For processes that exhibit common cause variability, the next step is to assess the KPOV relative to the needs of customers. This assessment is most typically made relative to specification requirements—for example, process capability/performance indices such as C_p, C_{pk}, P_p, and P_{pk}. However, we need to note that indices of this type can be deceiving, as described within Breyfogle (1999), *Implementing Six Sigma* (chapter 11).

A probability plot of the data can be a good supplementary tool in better understanding expected variability. This form of output can also be very beneficial in describing the "process capability/performance" of a transactional/service process. A probability plot for this process might indicate that 80% of the time it takes between two days and six weeks to fulfill orders. For an S^4 project in this area, someone might then estimate the cost impact of this process variability on an organization and/or on general customer dissatisfaction level. These values could then be the baseline against which S^4 projects are measured.

When this measurement and process improvement strategy is implemented, true change can be detected as a shift of the "30,000-foot level" control chart. This chart should be constantly monitored with an eye toward detecting a shift in the control chart pattern as a result of S^4 improvements.

4.3   DISCUSSION OF PROCESS CONTROL CHARTING AT THE "30,000-FOOT LEVEL"

We believe that "30,000-foot level" control charting should be an integral part of any Six Sigma strategy; however, some readers may feel uncomfortable with this approach, since it differs fundamentally from traditional control charting techniques. In an attempt to allay this concern, this section will elaborate on the technique further.

The classic approach to creating control charts is that subgroups be chosen so that opportunities for variation among the units within a subgroup are minimized. The thinking is that if variation within a subgroup represents the piece-to-piece variability over a very short period of time, then any unusual variation between subgroups would reflect changes in the process that should be investigated.

Let us consider a process that has one operator per shift and batch-to-batch raw material changes that occur daily. Consider also that there are some slight operator-to-operator differences and raw material differences from batch to batch, but that raw material is always within specification limits. If a control chart is established where five pieces are taken in a row for each shift, the variability used to calculate $\bar{x}$ and R control chart limits does not consider operator-to-operator and batch-to-batch raw material variability. If the variability between operators and batch-to-batch is large relative to five pieces in a row, the process could appear to be going out of control a lot. When a process goes out of control, we are told to "stop the presses" and fix the special cause problem. Much frustration can occur in the manufacturing process, because employees will probably be trying to fix a special cause problem over which they have no control.

One question that might be asked is whether the manufacturing line should be shut down because of such a special cause. It could even be questioned whether out-of-control conditions caused by raw material should be classified as special cause. Note that this point does not say that raw material is not a problem to the process even though it is within tolerance. The point is whether the variability between raw material should be treated as a special cause. It seems that there is a very good argument to treat this type of variability as common cause. If this is the case, control limits should then be created to include this variability.

To address this problem, we suggest an infrequent sampling plan where only one KPOV is tracked on a control chart. To address whether different variability sources are causing a problem, the long-term variability of the process could then be compared to specification needs, as illustrated in Figure 4.3. Mathematically, this comparison could be expressed in process capability/performance units such as C_p, C_{pk}, P_p, and P_{pk}. In addition, this variability could be described as percent beyond specification, or in parts-per-million (ppm) defect rates. If there is no specification, as might be the case in transactional or service processes, variability could be expressed in percent of occurrence through the use of normal probability plots. An example of this application occurs when 80% of the time a purchase order takes between 10 days and 50 days to fill.

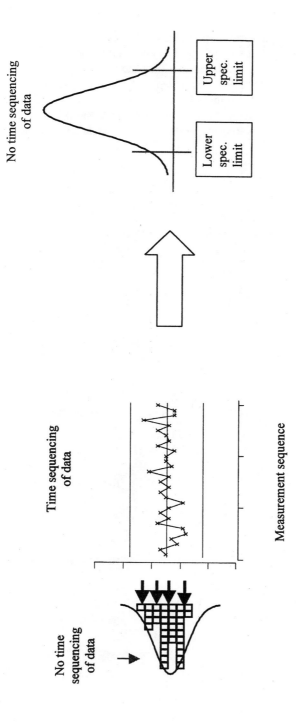

Figure 4.3 Control charting and process capability/performance assessment at the "30,000-foot level" view

If the process is not capable of meeting specification needs, then further analysis of the process using additional tools could show that the process is not robust to raw material batch-to-batch variations, as illustrated in Figure 4.4. Graphical tools described in Breyfogle (1999), *Implementing Six Sigma* (chapter 15), that give insight into these differences include multi-vari charts, box plots, and marginal plots. Analytical tools that mathematically examine the significance of these observed differences include variance components analysis, analysis of variance, and analysis of means.

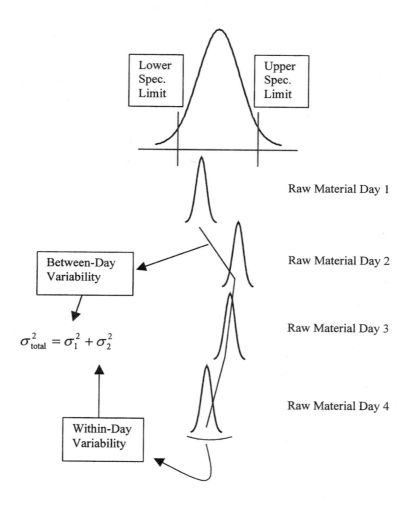

Figure 4.4   Potential source for output variability

When we react to the "problem of the day," we are not structurally assessing this type of problem. With an S^4 strategy we integrate structural brainstorming techniques (e.g., process flowcharting, cause-and-effect diagram, and FMEA), passive analyses (e.g., multi-vari analysis, ANOVA, regression analysis), proactive experimentation (e.g., fractional factorial Design of Experiments), and control (e.g., error proofing and control charting at the "50-foot level") to identify and fix the process problem and prevent its recurrence.

It should be noted that the discussion above does not suggest that an infrequent sampling plan using individual charting is the best approach for all processes. What it does suggest is that perhaps we often need to step back and look at the purpose and value of how many control charts are being used. Surely control charts are beneficial in indicating when a process needs to be adjusted before a lot of scrap is produced. However, control charts can also be very beneficial at a higher plane of vision that reduces day-to-day firefighting of common cause problems as though they were special cause.

4.4   CONTROL CHARTS AT THE "30,000-FOOT LEVEL": ATTRIBUTE RESPONSE

The preceding section described a continuous response situation. Let us now consider an attribute response situation. Care must be exercised when creating a "30,000-foot level" control chart. A control chart methodology should be considered whereby control limits that address variability between subgroups are emphasized. Breyfogle (1999, chapter 10), in *Implementing Six Sigma*, describes how an *XmR* control chart can often be more appropriate than a traditional *p* chart for attribute data analysis.

Figure 4.5 shows a situation where the capability of an attribute process (e.g., printed circuit board defects) is considered unsatisfactory. Our best estimate for the "30,000-foot level" capability is the mean of the control chart, which could be expressed in ppm units. The Pareto chart shown within this figure can give an indicator of where we should focus our efforts for improvement. However, we should exercise care when making decisions from such graphical tools. Measures such as the chi-square statistical test should be used to test for statistical significance.

4.5   GOAL SETTING, SCORECARD, AND MEASUREMENTS

We are trained to think that goals are very important, and they are. Goals should be SMART (i.e., *s*imple, *m*easurable, *a*greed to, *r*easonable, and *t*ime-based). However, these guidelines are often violated. Arbitrary goals that are set for individuals or organizations can be counterproductive and costly when indi-

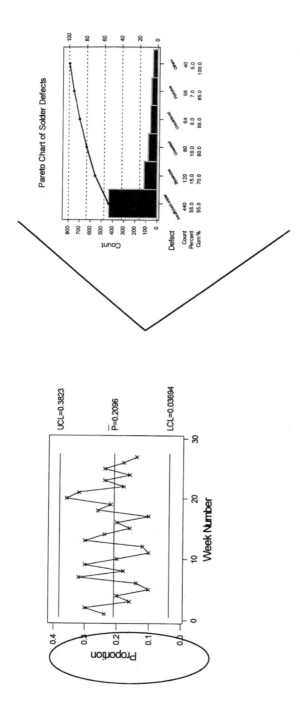

Figure 4.5 Control charts at the "30,000-foot level": Attribute response

66

viduals or organizations have no control over the outcome. Consider the following two situations within your personal life.

Situation 1

You have a lot of work that needs to get done around your house or apartment. Setting a goal to get these tasks out of the way on a Saturday would solve the problem. Consider these two alternative scenarios:

- You can set a goal for the number of tasks you would like to accomplish, create a plan on how to accomplish these tasks most efficiently, and track how well you are doing during the day to meet your goal. You would probably get more done with this strategy than by just randomly attacking a list of to-do items.

- Your spouse or friend can set an arbitrarily goal for the number of tasks you are to accomplish. If they don't create a plan and the goals are set arbitrarily, you probably will not accomplish as much as you would under the first scenario, since there is no plan and you do not necessarily have buy-in to the tasks that are to be accomplished.

Situation 2

You would like to make an overall rate of return of 50% on your financial investments next year. Consider these two alternative scenarios:

- You purchase mutual funds that have tracked well over the last few years. However, their yield to date is not as large as you had hoped. Because this goal is important to you, you decide to track the yield of your portfolio daily and switch money in and out of various funds in order to meet your goal.

- You evaluate the performance of a wide range of mutual funds that have multiyear performance records. You determine that your initial goal was too aggressive since the majority of funds have not produced the yield that you desire. Furthermore, funds that yield high rates of return one year often have trouble repeating this level of performance. You alter your plan and choose an investment strategy of buying and holding a diversified portfolio of index mutual funds.

Situations one and two are very different relative to goal setting. In the first situation, goal setting can be very useful when there is buy-in to the plan, because we do have some control over the outcome.

In the second situation, goal setting could be very counterproductive since we do not directly control the outcome of the process (i.e., how well the stock market will do next year). The first scenario in situation 2 will probably add variability to your investment process and may cause significant losses. The second plan for addressing this goal is the best, because we did research in choosing the best process for our situation and took the resulting yield, even though we might not achieve our goal.

Management can easily fall into the trap of setting arbitrary goals where the outcome is beyond the employee's control. Consider a manufacturing process that derives 90% of its problems from supplier quality issues. Often the approach in this situation is to "beat up" on suppliers whenever quality problems occur or to add inspection steps. This may be the only course of action manufacturing employees can take, but it will not be very conducive to meeting improvement goals.

However, if management extends the scope of what can be done by employees to fix problems, real productivity improvement can result. For example, manufacturing employees can become more involved with the initial supplier selection. Or perhaps manufacturing employees can be permitted to work with engineering to conduct a design of experiments (DOE) where several two-level factors are evaluated within an experiment for the purpose of improving the settings in the manufacturing process. Through this DOE, perhaps a change in a process temperature or pressure setting can be made that will result in a higher-quality product using raw material of lesser quality.

We do not mean to imply that stretch goals are not useful, because they are. Stretch goals can get people thinking "out of the box." However, goals alone—without a real willingness to change and a road map to conduct change—can be detrimental.

There is another consideration when setting goals: the scorecard. Vince Lombardi, the highly successful American football coach of the Green Bay Packers, said that "if you are not keeping score, you are just practicing." Organizations should not just measure against sigma quality levels and/or process output defect rates. If they do, they can be trapped into "declaring victory" with one metric at the expense of another metric. For example, the financial measurements of Xerox in the 1970s were excellent in part due to their servicing of the copiers that they developed and manufactured. However, customer satisfaction was low because of frequent paper jams and servicing. When less expensive and more reliable copiers were made available by the competition, the company had a severe financial crisis. In a later chapter we discuss considerations for a balanced scorecard.

The success of Six Sigma depends upon how goals are set and any boundaries that are set relative to meeting these objectives. In addition, if the scorecard is not balanced, people can be driving their activities toward a metric goal in one area while adversely affecting another metric or area. The result is that even though an employee may appear successful since he has met his goal, the overall organization can suffer.

4.6 SUMMARY

When dealing with numbers, information, and knowledge, we need a strategy that quantifies and tracks measurements that are vital to the success of the organization. A wisely applied Six Sigma strategy can facilitate productive utilization of information. Six Sigma techniques are applicable to processes as diverse as manufacturing, services, transactional, and DFSS (Design For Six Sigma). Pictures can provide powerful insights into the source of problems, where we should focus our efforts, and how we might efficiently go about improving products and processes. However, we must never forget the importance of data analysis and statistical assessments in reducing the risk of reacting to visual observation of problems that could have occurred by chance.

5

CRAFTING INSIGHTFUL METRICS

5.1 SIX SIGMA METRICS

Metrics can be very beneficial within an organization when they are used in such a way that they give direction to what is the most appropriate activity for a given situation. However, some metrics can have no value or might even counter productive activity. For example, little value is realized when an organization reports metrics merely as a status and never makes an orchestrated effort to improve them. In addition, counterproductive results can occur when organizations choose to use one type of metric throughout their organization. In other words, creating a "one-size-fits-all" metric, such as sigma quality level, as described earlier in this book, can lead to frustration, fabrication of measurements, and "playing games with the numbers."

Organizations should consider using only those Six Sigma metrics that are appropriate for each situation. Chapter 9 of Breyfogle (1999), *Implementing Six Sigma*, not only describes many possible Six Sigma metrics alternatives—including DPMO, rolled throughput yield, and process capability/performance (C_p, C_{pk}, P_p, and P_{pk})—but also addresses the confusing and controversial aspects of metrics. Next we will describe some of these Six Sigma metrics followed by example strategies.

Process Capability

Process capability is a commonly used metric within Six Sigma. The following equations are used to determine the process capability indices C_p and C_{pk}:

$$C_p = \frac{USL - LSL}{6\,\sigma} \quad \text{and} \quad C_{pk} = \min\left[\frac{USL - \mu}{3\,\sigma}, \; \frac{\mu - LSL}{3\,\sigma}\right]$$

where USL and LSL are the upper specification limit and lower specification limit, respectively, σ is standard deviation, and μ is the mean of the process.

Figure 5.1 and Figure 5.2 illustrate how the distribution of a dimension could be related to a specification, yielding various process capability indices of C_p and C_{pk}. A process is said to be equal to at "six sigma quality level" if $C_p = 2.00$ and $C_{pk} = 1.5$, which equate to 3.4 PPM using the 1.5 standard shift, as previously described within this book.

For a visual image of the meaning inherent to C_p, consider that the numerator of the C_p equation represents the width of a garage door opening. Likewise, consider the denominator of the C_p equation to be the width of a new sports car. If the door width and the car width are identical, then C_p is equal to a numerical value of 1. It also means that the driver needs to be extremely careful when pulling into the garage. Any slight miscalculation will lead to a scratch or dent. A process with a C_p value of one is not considered capable of consistently meeting specifications, because any slight shift in the process mean, or any increase in process variation, would lead to the production of defective product based on a $\pm 3\sigma$ processes requirement.

A C_p value of two means that the garage door is twice as wide as the car, or 12σ units wide. A process capability index of 2 means that even if the process

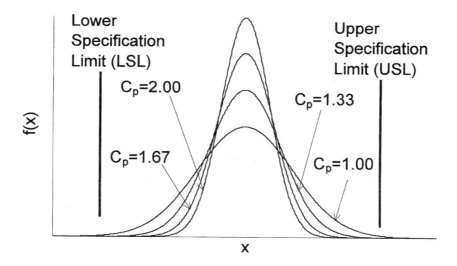

Figure 5.1 C_p Examples

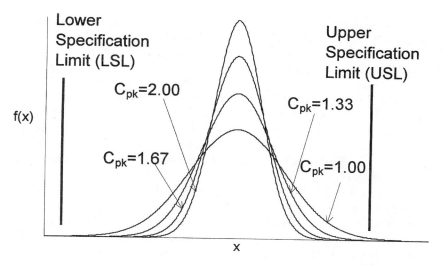

Figure 5.2 C_{pk} Examples

variability were to double to an exact width of 12σ units, the process still would not exceed the customer's specification limits if it were centered.

Care must be exercised when determining and utilizing C_p and C_{pk} process capability indices. First, the previously described equations are appropriate when data are normally distributed. Second, the sample size and how the data are collected from a process can dramatically affect estimates of these σ values. If data are collected infrequently from a process and then analyzed, we might get a value for process capability data dramatically different from that gotten when the data are collected over a very short period of time.

In general, there are several methodologies for data collection and procedures for calculating process capability values that can result in very different results for C_p and C_{pk} values. Breyfogle (1999), in *Implementing Six Sigma* (chapter 11), describes many procedures for the calculation of standard deviation for process capability/performance equations. These procedures vary from determining a short-term value to determining a long-term value using variance components techniques. Some people insist that a short-term view of standard deviation is to be used with respect to these equations; others hold that a long-term view of standard deviation is proper. Obviously, these methodologies can give very different results.

Another problem with requiring that all organizations use process capability metrics within Six Sigma is that typically service/transactional processes do not have published specifications that are similar to those within manufacturing. To calculate C_p and C_{pk}, specifications are needed; hence, people could be forced to fabricate specifications in order to determine process capability indices for this situation. This activity is non-value added.

As an alternative to C_p and C_{pk} we prefer tracking long-term estimates of the defect rates or the percent of product outside of the specification limits. For service/transactional processes, including cycle times, we can describe estimates and report, for example, what we expect to occur 80% of the time. The financial implication for both these situations could then be estimated and compared to other areas of the business for the purpose of determining the amount of effort that should be given to process improvement for these areas of the business. A financial return could also be estimated for these opportunities for improvement.

To avoid confusion, we suggest using "hard" and "soft" money as a primary measure when communicating the value of a S[4] project. In its 1997 and 1998 letters to shareholders and employees, GE either directly or indirectly implies the monetary benefits of Six Sigma for the described Six Sigma projects. In these reports, no mention was made of the other metrics often associated with Six Sigma. This does not mean to imply that other Six Sigma metrics should not be used within a project (or be the primary driver of a project); however, it does suggest that we need to select metrics with care, and not overlook a metric that executives and others understand the most: money.

Process Cycle Time

During the past decade, cycle time has become an increasingly important competitive metric. Most people simply assume that quicker means better, but that is not always the case. The advantage of shorter cycle times is much more fundamental, and it relates to innovation and technology maturity.

Theoretical process cycle time is simply the ratio of the true amount of time it takes to create a batch of product with no wasted time, materials, or effort, divided by the number of units in the batch. By comparison, *real* process cycle time includes not only the time required to manufacture the product and its various components but also the time to inspect, ship, test, analyze, repair, set up, store, and so forth. The closer we can drive real cycle time toward its theoretical value, the more efficient we are at generating profit. We can obtain better delivery, less cash in inventory, and increased manufacturing capacity.

There is another advantage to shorter cycle times. If a product development cycle currently takes three years of effort, it means that the product shipped to customers on January 1, 2000, contains technology that was current as of January 1997. If a competitor has a much shorter—say, six-month—product development cycle for the same class of product, then their January 1, 2000, release contains technology that can be two and a half years more current. The strategic value of shorter cycle times is technological superiority.

DPU and DPMO

The defects-per-opportunity (DPO) calculation, which includes the number of opportunities for failure, can give additional insight into a process. For example, if there were 100 circuit boards and 5 of them failed final testing, there would be a DPU rate of 5%. However, if each one of the 100 boards had 100 opportunities for failure, and a total 21 defects were found when manufacturing the 100 circuit boards, the DPO rate would be a much smaller 0.21%. (21 defects per 10,000 opportunities). The DPO rate is often expressed in terms of defects per million opportunities (DPMO). In the previous example, a 0.0021 DPO rate equates to a 2,100 DPMO rate.

A DPMO rate can be a useful bridge from product metrics to process metrics. For example, within a printed circuit board manufacturing facility there can be many different printed circuit board products that pass through a manufacturing line. Each of these products can have different complexity and opportunities for failure when exposed to process steps such as pick-and-place of components and wave soldering. Typically, for this situation the opportunities for failure are the number of components plus the number of solder joints. For one day of activity, we could determine the total number of failures and divide this value by the number of opportunities to get a DPMO rate. This could be done each day, even if there were different product types produced daily. This metric could be tracked as a failure rate on a control chart using daily time increments. A Pareto chart could then be used to quantify the frequency of failure types and give direction to process improvement efforts.

Table 2.1 can be used to convert a DPMO rate from a process into a sigma quality level metric; however, for the reasons noted in Chapter 2, we do not recommend doing this. Instead, we recommend leaving the unit of measure as DPMO and quantifying the importance of reducing this defect rate through the conversion of the metric into its monetary implications.

Advice for Using DPMO and Sigma Quality Level

Some organizations require that a sigma quality level metric be reported for all divisions. Unfortunately, this requirement can lead to "playing games with the numbers." To illustrate this point, let us consider two types of situations that an organization can encounter.

The previously described DPMO methodology of tracking a process failure rate within a printed circuit board manufacturing facility made sense; however, in other situations it does not. For example, let us consider how a manufacturer of hot-water heaters might approach this situation if they were required to report a sigma quality level. For daily production they could determine the number of failures; however, the question of concern is quantifying the number of opportunities for failure. If we were to merely count the number of components that were parts of the hot-water heater assembly process, we may get a value of only 4 parts. If we were to calculate a DPMO rate using this number (i.e., 4) for failure opportunities, our sigma quality level would be very poor. In an effort to make our numbers look better, we might get ridiculous and "count the number of 0.001 inches of welding joints within the hot-water heater assembly process" as opportunities for failure. If we were to do this, our DPMO rate could perhaps improve such that a 25% yield (75% defective rate) appears to have a seven sigma quality level. We don't mean to suggest or condone such activity when calculating a Six Sigma metric; however, this is the type of activity that actually occurs whenever an organization requires a metric such as sigma quality level across all organizations and processes.

Another situation that can also lead to "playing games with the numbers" occurs if a company requires sigma quality level measures for its service/transactional processes. For this situation we might have basic desires of customers but no specification requirements as they would typically exist within manufacturing. For example, within a service/transactional process someone might have chosen a target that all deliveries are to be within ±1 day of their due date; however, this specification does not have the same importance as a tolerance for a piston that is to be part of an automobile engine assembly. When we calculate many of the Six Sigma metrics, such as process capability/performance, DPMO rates, and sigma quality levels, we need specification values in order to perform the calculations. If we are to force any of these Six Sigma metrics—for example, sigma quality level within service/transactional processes—the organization responsible will often need to fabricate specification values in order to perform the calculations. This activity can lead to the adjustment of these "specification values" so that their overall sigma quality value "looks good."

Hopefully, our hot-water heater and service/transactional examples illustrate the dangers that can occur when an organization succumbs to the temptation of using one-size-fits-all metrics across organizational disciplines. The metrics associated with Six Sigma can be very valuable; however, it is important to determine, use, and report the most appropriate metric(s) for any given situation/process. In addition to reporting and tracking the important metrics within a process, we suggest also reporting the monetary impact of the metrics.

Yield and Rolled Throughput Yield (RTY)

When organizations report the final yield of their process, they may be referring to the percentage of units that pass final test relative to the number of units that entered the process. This measurement can fail to quantify the amount of rework that can occur within a process before final test, where this rework is called the "hidden factory." The hidden factory does not occur only within manufacturing; it also occurs within business processes. Rolled throughput yield is a metric that can expose the location and magnitude of the hidden factory.

Figure 5.3 graphically illustrates the difference between process yield and RTY. We will use the left side of this figure to illustrate RTY and the right side of this figure to illustrate process yield. In both cases we will assume that our master schedule calls for us to produce the "1,000 units" highlighted at the top of the figure. Notice that our process consists of five subprocessing steps, each with its own distinctive yield (Y_i). The yield of the first subprocess, Y_1, is 0.92, or 92%. Focusing again on the left-hand side of this figure, we notice the number of units decreasing from 1,000 to 920 to 754 to 633 to 519 to 493. This is a direct result of applying each subprocessing step's yield directly to the number of units that initially enter the work cell for batch processing. When 1,000 units enter subprocess Y_1 only 920 units exit without some form of further processing. Looking at the right side of the figure, one can tell where those 80 units (1,000 – 920) went: 40 were scrapped and 40 were reworked and put back into the production cycle at some later time. For this simple example we will assume that all reworked material successfully passes to the next subprocess without further incident and that the percent reworked and scrapped are the same at each process step. The 920 units that avoided scrap and rework during subprocess Y_1 enter the second subprocess, Y_2 (with a measured yield of 82%), resulting in 754 good parts. And so on until all five subprocessing steps deliver 493 good units. This 49.3% rolled throughput yield (RTY) value can also be estimated by merely multiplying the yields of the five sequential subprocessing steps. The numbers are identical when the round-off error of the numbers is ignored.

The right side of Figure 5.3 illustrates how process yield can be calculated. During the first subprocess the initial 1,000 units is reduced to 960, since we had to scrap 40 units that could not be salvaged. The reworked units (4%, or 40 units) subjected to rework will eventually show up in inventory and hence are counted as good product. The only deviation from perfection is the scrap (40 units), resulting in a process yield through subprocess Y_1 of 96%, or 960 units. By the time we get to the end of the fifth subprocess, the process yield has declined to 71.4%.

The two methods of calculating yield for this example are 71.4% and the 49.3%. The accounting systems of organizations typically ignore rework and report the higher value. The rework process that we are missing within our calculations is referred to as the "hidden factory."

In summary, rolled throughput yield is a measure of how well companies "process quality." Process yield measures how well one "processes units." It is not uncommon to have a multistage process where yield at each step is quite high (say, at or above 90%) but the rolled throughput yield is at 50% or below. An RTY of 50% indicates that only one out of two units completes the entire production process without being reworked or scrapped, giving more insight into process performance.

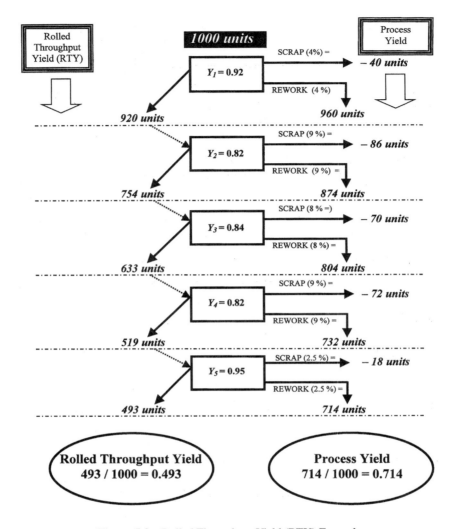

Figure 5.3 Rolled Throughput Yield (RTY) Example

5.2 RIGHT QUESTION, RIGHT METRIC, RIGHT ACTIVITY

Executive direction can sometimes lead to the creation and mandatory use of metrics that have no long-term or even short-term benefit. For example, the vice president of manufacturing might indicate that she wants a weekly listing of defectives produced on the ABD production line, and a subsequent action plan for fixing these defectives so they don't recur. This type of direction often leads to firefighting, where common cause variation is treated as if it were special cause. A more appropriate question along these same lines might be: "Provide me a control chart showing the defect rate over time. If our process is in-control and the average failure rate is too high, provide a Pareto chart describing the frequency of defect types over several months and then describe an action plan to reduce the magnitude of the categories that have the largest value."

Rather than engaging in firefighting, management should consider asking for a "30,000-foot level" control chart of defectives, as illustrated in Figure 5.4 and as described previously in this book. The control chart on the left side of this figure shows an average defect rate slightly above 20%. This process is considered to be in statistical control, since there are no special causes present based on these data. Assuming that this reject rate is unacceptable, a reasonable next step is to organize observed defects into logical categories. A transformation of this data into a Pareto chart is also illustrated in Figure 5.4. A potential next step would be to request a project plan to address the most frequent category of defect, using Six Sigma tools. Also, it might be helpful to quantify how much each type of defect is costing the company in financial terms. This would help prioritize this potential Six Sigma project relative to other improvement opportunities.

The example used above involved percent defective attribute data. Figure 5.5 illustrates what a completed process improvement scenario might look like when plotting continuous data in the form of an individuals control chart. This control chart indicates that some sort of process change has created a significant shift in the process mean to a "better (smaller)" value. When we superimpose customer specification limits on the $\pm 3\sigma$ normal distribution, we get a visual representation indicating that the frequency of nonconformance has gone down significantly as a result of the process change.

These two examples show how the "30,000-foot level" control chart can be used to give meaning and direction within an organization. The other Six Sigma metrics described earlier in this chapter can be beneficial for tracking; however, they are not appropriate for every situation. In many cases their use can be a terrible waste of time and effort. For example, rolled throughput yield (RTY) is not something that should automatically be measured whenever there is a troublesome product or process. It takes a lot of time and money to collect the data necessary to estimate rolled throughput yield on a process of even a modest size (five to ten steps). At each individual step in the process one would need to keep track of scrap, rework, number of defects, number of opportunities for defects, and so forth. In addition, one needs to permit enough product to flow through the process so that reasonable estimates of throughput yield can be obtained for each process step. One would only go down this path for data collection when there is evidence that a process is producing a lot of rework and/or scrap and if this data collection can give insight to meaningful improvement/tracking activities.

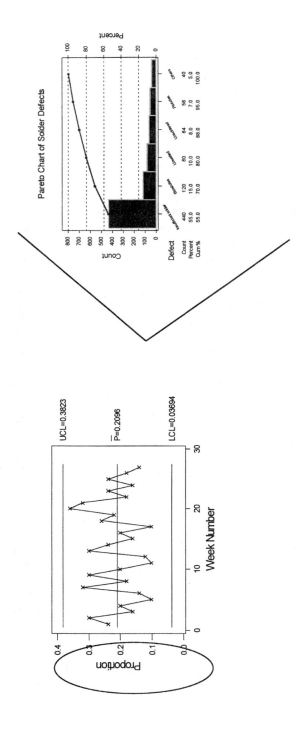

Figure 5.4 Control charts at the "30,000-foot-level": attribute response

A Key Process Output Variable

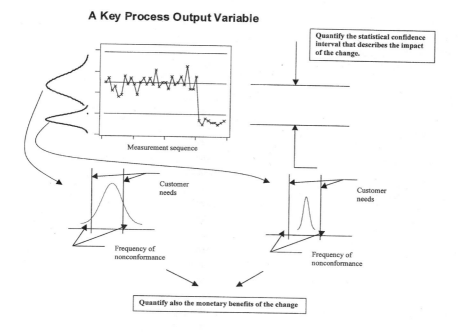

Figure 5.5 Tracking project metrics and quantifying improvements

5.3 EXAMPLE 5.1: TRACKING ONGOING PRODUCT COMPLIANCE FROM A PROCESS POINT OF VIEW

An organization is contractually required, per government regulation and law, to periodically sample product from a manufacturing line to assess ongoing compliance to mandated quality standards. One of the tests is time consuming and is known to have a lot of testing error, even though the amount of variability has not been quantified. Many government products manufactured by this company are subject to this testing procedure. They are all similar in nature but have different design requirements and suppliers. To address these government requirements, the organization periodically samples the manufacturing process and tests for compliance to specifications. If the product does not meet requirements, corrective action is taken within the manufacturing organization. The organization is concerned with the amount of resources spent in resolving these noncompliance issues.

When we step back and look at this situation, we observe that the current testing involves attribute pass/fail data where measurements are focused on the product, not the process. Also, when samples are outside specification limits, reaction to the situation is to immediately assume that the problem is the result of special cause variation (i.e., let manufacturing identify and fix the problem).

Let's look at the situation from a "30,000-foot level" perspective by examining all of the data as if it came from the same continuous process. We are going to ignore the individuality of the distinct product lines, since all products basically encounter the same process of design, design qualification testing, and manufacturing. In other words, we are considering product differences to be noise to our overall process. From this point of view, we can hope to determine if the failures are special or common cause in nature. If the failures are due to common cause variation, we might be able to collectively view the common cause data to gain insight to what should be done differently within the process to reduce the overall frequency of failure. If the failures are due to special cause variation, we might be able to examine this/these special cause condition(s) individually to determine why these data differ from our normal process and then take appropriate action(s).

In lieu of compiling and reporting the data in attribute format, we will collect and report conformance data in the form of continuous data. The raw data used to construct the *XmR* control charts in Figure 5.6 are listed below. The time sequence in which the data were collected can be reconstructed by reading the raw data from left to right, starting with the first row and continuing to rows two and three, using the same left-to-right pattern as in row one. The government specification simply states that raw data values at or below zero are all equally acceptable, whereas positive values are unacceptable, with the degree of unacceptability increasing as the positive values increase. In other words, 0 is good, 0.1 is bad, 1.0 is worse, 10.0 is terrible, and so forth. Any raw data values greater than zero for this sampled product represent an out-of-specification condition that must be resolved within the manufacturing line.

−9.4	−9.4	−6.4	−7.7	−9.7	−8.6	−4.7	−4.7	−6.2	−0.3	3.0	0.6
−8.8	−9.5	−2.6	−6.9	*11.2*	−9.5	−9.3	−7.8	−12.4	−3.2	−4.9	
−16.3	−3.5	−6.7	−10.0	3.6	−0.2	−7.6	1.9				

The *XmR* chart in Figure 5.6 indicates that the seventeenth sequential data point, the bold value of 11.2 in the raw data table, is "out of control" or statistically different from the other readings from the process. Our initial reaction to the magnitude of this point is that a special cause condition may now be occurring within the manufacturing process that needs to be resolved. It should be noted that the number(s) 1 plotted within Figure 5.6 are placed there by the statistical analysis software to indicate that rule number 1 has been violated. Rule number 1 indicates that the process is out of control because a single data point is outside of the ±3σ control limits. See Chapter 10 in Breyfogle (1999), *Implementing Six Sigma,* for additional information on these and other control chart rules.

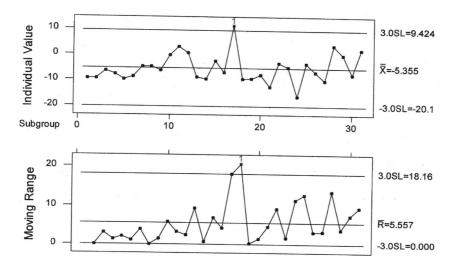

Figure 5.6   XmR chart of data

Figure 5.7 shows a normal probability plot of the same raw data. This plot indicates two important results: First, only a single data point in the upper right-hand corner of the normal probability plot seems to be inconsistent with the maximum-likelihood straight line that is determined by the statistical analysis software. From this, one might conclude that only one of the 31 data points appears to be an outlier. This conclusion is consistent with our results from the *XmR* charts. Second, if we look where the normal probability plot maximum likelihood regression line crosses the 0 point on the X-axis (83%), we estimate that 17% of product fails to meet specifications. This means our overall manufacturing process is not very capable of consistently meeting the customer specification requirements. In other words, we might consider that the outlier or special cause event is only part of our problem. We have a process that is not very capable and we need to make one or more *process changes* if we want to improve quality and reduce the amount of scrap and rework.

When we went to the manufacturing line, we identified that there was a special cause problem for the seventeenth data point from this data set, which later was resolved. So we need to reevaluate the process with this data point removed. The resulting normal probability plot for this adjusted set of data is shown in Figure 5.8. This plot shows a much better fit, where we would now conclude that about 10% of overall product from the manufacturing process is beyond the specification limit, whenever there are no special cause conditions.

We would next like to determine what improvement(s) should be made within our process. Using existing data, we could create a Pareto chart of what had

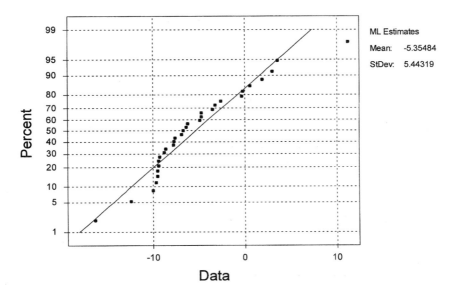

Figure 5.7 Normal probability plot of all data

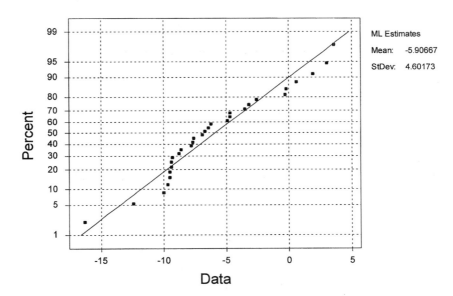

Figure 5.8 Normal probability plot of data without special cause condition

been done previously within manufacturing to bring out-of-specification product back into compliance. Perhaps historical "fixes" will point toward some "common part or processing step" among the different product types that would suggest a common solution. Furthermore, if we do identify a common solution to this problem we should pass the information along to both product development and manufacturing so that the design is more capable of consistently producing product that is within specification limits. We might also look for a poka-yoke (error-proofing) solution that would eliminate the out-of-specification condition during the manufacturing process.

So far we have looked at data in different ways without making any additional measurements. The next process improvement effort that seems appropriate is to better understand our measurement system through a Gage R&R study. When this is complete, a DOE may give insight into what could be done differently within the process.

5.4   EXAMPLE 5.2: TRACKING AND IMPROVING TIMES FOR CHANGE ORDERS

An organization would like for engineering change orders (ECO) to be resolved as quickly as possible. The time series data that follow indicate the number of days it has recently taken to complete change orders:

18	**2**	0	0	0	0	0	0	0	0	0	0	0
	0	0	0	0	0	14	0	0	7	3	0	41
	0	0	0	0	0	0	0	0	17	0	0	0
	0	0	0	0	0	0	0	0	0	0	0	0
	0	**1**	0	0	0	0	0	0	11	0	0	17
	26	0	0	0	0	0	21	0	0	0	0
	6	0	0	17	0	0	0	0	0	0	0	0

The *XmR* chart of this data shown in Figure 5.9 does not give much insight into the process, since there are so many zeros. We should note that an *XmR* chart was chosen to analyze this data so that between-sample variability would be considered within the calculation of the control limits, as described within Breyfogle (1999) in *Implementing Six Sigma* (chapter 10).

A histogram of this data indicates that it might be better to consider the situation as bimodal in nature. That is, we might consider "less than one day (i.e., 0)" as one distribution and "one day or more" as another distribution. An estimate of the proportion of ECOs that take "one day or more" to process is 14/85 = 0.16. As shown in Figure 5.10, the *XmR* control charts of the "one day or

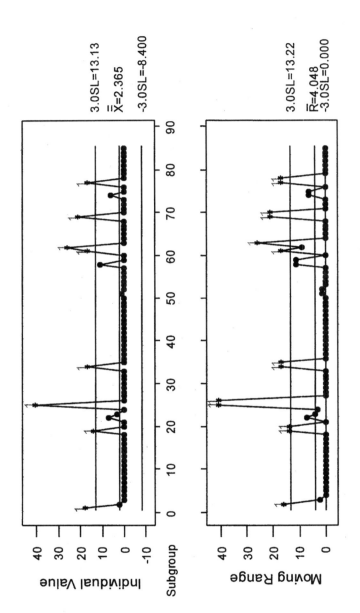

Figure 5.9 XmR chart of the time taken to process engineering change orders

more" values do not indicate any evidence of the existence of special causes of variation. The normal probability plot shown in Figure 5.11 is another way of viewing the data when ECOs "take one day or more" to process.

We can combine the information contained in these two figures to give an estimate of the percentage of change orders from the process that take longer than any specified time period of our choosing. To illustrate this, let's estimate what percentage of the time ECOs take 10 or more days to process. We have already estimated that the proportion of the ECOs that take "one day or more" to process is $14/85 = 0.16$. Since we are interested in a target value of 10 or more days, this ratio of 0.16 applies to our situation. We can use the normal probability plot in Figure 5.11 to estimate that the proportion of ECOs that take one day or more to process is 0.66, or 66%, since from the figure $1 - 0.34 = 0.66$. Therefore, the estimated percentage of engineering change orders taking 10 days or longer to process is $[0.16] \times [0.66] = 0.11$, or 11%.

With this information we could categorize the characteristics of change orders that take a long period of time. We could also create a Pareto chart that would give us a visual representation that leads to focus areas for process improvement efforts. Also, we could calculate the monetary implications of late change orders as they impact production and/or development deliveries.

5.5   EXAMPLE 5.3: IMPROVING THE EFFECTIVENESS OF EMPLOYEE OPINION SURVEYS

Many employee opinion surveys are conducted periodically—for example, annually or semiannually. These surveys typically involve the use of a Likert scale with associated numerical responses, such as the following:

Strongly agree—5

Agree—4

Uncertain—3

Disagree—2

Strongly disagree—1

Much time and effort can be spent trying to understand the meaning of the results. And even after considerable analysis, we might question the value of the survey. When interpreting the results from the survey, for example, one might ask several questions:

- Is the rating scale appropriate for what we are trying to accomplish?
- Are there any significant positive or negative trends evidenced by the data?

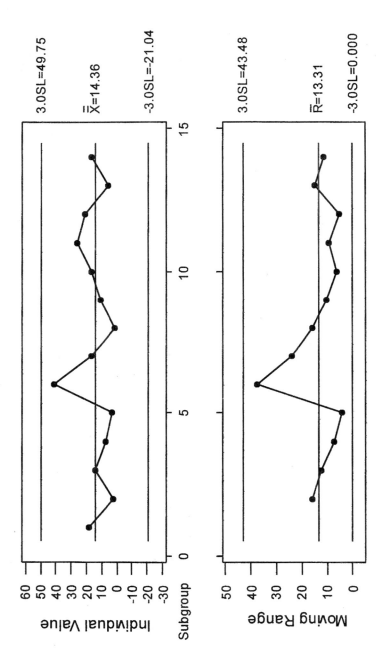

Figure 5.10 XmR chart of nonzero values

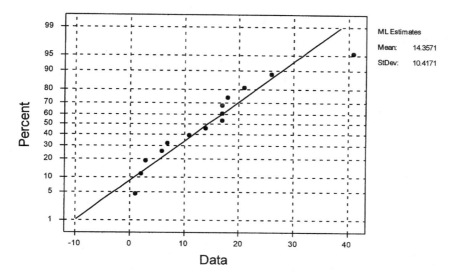

Figure 5.11 Normal probability plot of nonzero values

- Might the results be different if the survey were taken at a different time—for example, right after the release of performance appraisal data to employees, after the U.S. Internal Revenue Service income tax filing deadline of April 15, during the Christmas holidays, and so forth?
- What could or should be done differently during the next survey cycle?

Let us examine what might be done differently so that survey design and analysis will focus on gaining insight into company internal processes—as opposed, for example, to ranking individuals and/or departments and then taking "appropriate corrective action as required."

The use of the Likert scale with integer responses on a 1-to-5 rating scale and with distinct upper and lower bounds can lead to interpretation problems. For example, if most or all employees rate one or more of the survey questions at a 5 level (i.e., strongly agree), it will be virtually impossible to measure subsequent improvement. This issue could be addressed by asking respondents to answer the question(s) relative to changes from the previous year. This would involve modifying the rating scale to:

Significant improvement—5

Some improvement—4

No change—3

Some decline—2

Significant decline—1

Since there are no upper bounds with this approach, there is much more flexibility. For example, management could set a stretch goal, consistent from year to year, of 3.5 or greater average response, with 80% believing improvement has not taken a turn for the worse, which equates to a score less than 3.0. Because of the phrasing of the question, only those who had been employed for one year or more could serve as respondents. This might require the addition of a sixth response category, such as "not employed last year" or "not applicable."

To address the bias in the results due to "noise" such as weather and fear of layoff, we could randomly divide the workforce into 12 separate groups and assign each group to be surveyed in a different month of the year. A control chart could then be used to evaluate overall and specific question responses. Other Six Sigma tools, such as normal probability plotting and analysis of means (ANOM), can give insight into variability and other areas.

Valuable information for improvement opportunities is often contained in the "comments" section of the survey. However, it can be very difficult, if not impossible, to classify and rank-order these suggested improvements without also being able to ask respondents how much they value the proposed changes. To address this concern, consider adding a section that lists improvement ideas. Each person can vote on a certain number of listed ideas. This information can then be presented in a Pareto chart format. Ideas can come from previous years' write-in comments or they can be created ahead of time by a designated committee of cross-functional experts presumed to be sensitive to employee needs and concerns.

Although it makes survey design, execution, and analysis considerably more difficult, consider having different lists of improvement ideas for different groups within the organization. For example, the administrative staff may be highly satisfied with their computer workstations, whereas the engineering group is clamoring for a supercomputer to conduct complex finite-element analysis and three-dimensional modeling of complex structures. A single question on "ready access to required information technologies" could lead to considerable confusion when attempting to explain strong differences of opinion between the two groups of workers.

5.6   EXAMPLE 5.4: TRACKING AND REDUCING OVERDUE ACCOUNTS PAYABLE

Invoices were selected randomly over time from an accounts receivable department. Figure 5.12 shows a histogram of the number of days that these invoices were overdue in payment. Negative numbers indicate that payments were made before the due date. It is quite evident that the data are not normally distributed, since the distribution is skewed to the right. We could try to use a three-parameter Weibull distribution or Box-Cox Transformation (Box et al., 1978) to better describe the data. It should be noted that for this study the measurement was changed from attribute to a continuous measurement relative to due date. Continuous response data are preferable to attribute data, such as a yes or no response, that address whether invoices have been paid on time or not. With continuous data we can obtain more information about a process with less data. In addition, we can quantify the amount of time overdue accounts are delinquent, which can yield a better monetary impact for the current situation.

When collecting the "days overdue" data, other potentially useful variables were simultaneously recorded. Variables that were collected for this example included company invoiced, amount of invoice, whether the invoice had errors upon submittal, and whether any payment incentives were offered to the customer. This supplemental data permits other kinds of useful analyses. For example, by knowing the monetary amount of the invoice and the days overdue, it was possible to calculate the time value of money the company was forgoing by not being paid on time. This number could then be used as a COPQ baseline for justifying an S^4 project.

Analysis of variance (ANOVA) and analysis of means (ANOM) studies indicated that the monetary amount of the invoice did not significantly affect the delinquency. It did reveal, however, that certain specific companies were consistently more delinquent than others. Also, invoices were frequently rejected

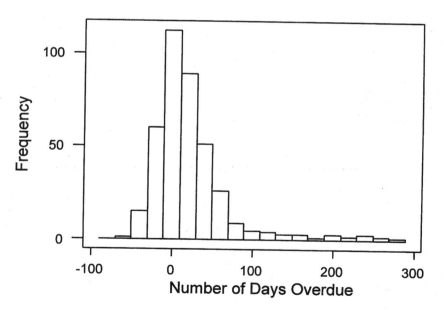

Figure 5.12   Histogram showing the number of days invoices are delinquent

because of errors detected upon receipt by the customer, which caused payment delays until the errors could be corrected. ANOVA and ANOM techniques are discussed in chapters 24 and 25 of Breyfogle (1999), *Implementing Six Sigma.*

In addition, analysis showed that the offer of a significant discount for timely payment did not significantly affect the likelihood of prompt or early payment. The companies taking advantage of the discount would probably have paid the invoice on time even without the discount. So, because the discount was not an effective incentive and eliminating it would probably not reduce the rate of delinquency, it was therefore deemed an unnecessary expense whose elimination would improve the bottom line of the company.

The collection of this additional data permitted extremely useful—and inexpensive—post hoc analyses of overdue payment data. With this additional insight, the team was able to quickly modify the accounts payable billing and collection process into a more effective one.

6

PERFORMANCE MEASUREMENT

In recent years, much has been written about performance and performance measurement. Increasingly, organizations are hiring consulting firms to help with these performance measures. Six Sigma project selection should take these metrics into account. However, we must keep in mind that some performance measures can lead to short-term activities that are not beneficial. In this chapter we discuss measurement types, the principle of measurement, and the balanced scorecard. These topics can be beneficial when executing projects.

6.1 MEASUREMENT TYPES

Thor (1988) has found that there are three main reasons for measuring things in organizations: to plan, to screen, and to control. Planning measures, normally the domain of senior executives, attempt to answer the question "Are we achieving our long-term strategic goals?" They are typically expressed in the language of executives, which is money and ROI. Therefore, it is appropriate that one Six Sigma category of measurement is planning. Because planning measures focus on long-range performance issues that tend to change slowly, they may be tabulated infrequently, perhaps once or twice a year.

Screening measures address the question "Are the functional areas performing in ways that support the organization's strategic goals?" They are expressed in both monetary and nonmonetary terms. Their focus tends to be on short- to intermediate-range performance; hence, they may be measured monthly or quarterly. Screening measures assess the performance of an organization's functional areas. For example, screening measures in a manufacturing organization would track performance in the areas of operations, sales, marketing, shipping, inventory control, research and development, and so on.

Control measures are probably the most familiar category, since they have received considerable attention from quality professionals. These measures address the question "Are employees, machines, products, services and processes performing in ways that are aligned with organizational and functional goals?" These measures are generally expressed in nonmonetary terms. They focus on immediate performance issues, and as a result, are measured daily, in some cases every few hours or minutes. Control measures assess the performance of individual people and things. We believe that the widespread resistance among workers (Bouckaert and Balk, 1991; Lobbestael and Vasquez, 1991) to measurement arises because workers see any measurement initiative as an attempt by management to exert control. Measurement efforts are probably a waste of time if their only purpose is to enable executives to control employees' behavior (Thor, 1989).

Planning and screening measures by their very nature represent less threatening forms of management intervention, and consequently, these measures should be considered in a Six Sigma business strategy.

The American Productivity and Quality Center teaches that there are a minimum of six elements required for a TQM initiative to be successful:

Quality planning
Team processes
Education and training
Recognition and reward
Communications
Measurement

Organizations that have found a previous TQM effort to be ineffective might have been missing one or more of these systems. A Six Sigma business strategy should encompass the systems discussed above, relying heavily on metrics that are used wisely and create insight rather than "playing games with the numbers."

6.2   PRINCIPLES OF MEASUREMENT

There are a number of fundamental principles associated with effective performance measurement. Among them are the following:

- Know why measurements are made and how they will be used
- Measure only what is important

- Measure causes (drivers) of good performance
- Use a family of measures
- Measure both internal and external views of performance
- Keep the number of measures small
- Provide feedback to those providing "performance data"
- Adjust monetary measures for inflation

Know why measurements are made. The use of measures to control machines, processes, and products, for example, can be beneficial. However, performance measures to control individuals can be counterproductive. It should be noted that control measures are intended to "lock in" the quality gains that have been achieved following changes in a product or process. We suggest using high-level performance not as a control metric but as one that helps with project selection and the wise implementation of Six Sigma.

Measure only what is important. We say that financial metrics should be considered as part of the infrastructure when selecting Six Sigma projects. The success of S^4 projects relative to financial measures can breed upon itself in such a way that more S^4 projects are created, leading to additional successes. Also, excessive measurements can be expensive—both in the time and money needed to collect data, and in the forgone resources when the data is responded to in an inefficient manner. To address this situation, we suggest giving emphasis to identifying and tracking only KPIVs and KPOVs for processes. GE made reference to these terms as X's (i.e., KPIVs) and critical-to-quality (CTQ) characteristics, or Y's (i.e., KPOVs), where CTQs are the items essential for customer satisfaction in a product or service and KPIVs are the key drivers.

Measure causes (drivers) of good performance. Historically the most important metrics have been financial measures such as return on investment (ROI), net profit, price to earnings (P/E) ratio, and stock price. These measures are outcomes, not drivers; they are effects, not causes. Six Sigma can guide organizations through the determination, measurement, and improvement of the response of KPOVs that affect financial measures. Six Sigma can then give direction on the determination and control of KPIVs that affect these KPOVs.

Use a family of measures. The "family of measures" concept is simply the belief that performance is multidimensional. Most processes worth measuring are too complex to be assessed using a single variable. A single measure of success can lead to the optimizing of one factor at the expense of others. The American Productivity and Quality Center (1988) asserts that measuring the three variables of productivity, quality, and timeliness captures the essence of group performance for knowledge workers. If multiple performance measures are chosen with skill, any false improvement recorded as a result of focusing on only one of the variables will be neutralized or minimized by a corresponding loss of achievement in a different variable from the same family.

Measure both internal and external views of performance. A good performance measurement system addresses both external and internal views of quality. Neither internal nor external measures are inherently bad. The problem comes when one chooses to emphasize one at the expense of the other. If emphasis is given only to internal measures of performance, the voice of the customer is lost. It is also easy to miss what is important to the customer if no effort is made to ask. Likewise, if emphasis is given only to external measures of performance, the significance of the internal costs of an organization may be overlooked. For example, an organization may produce a very-high-quality product that the customer loves, but in order to reach that quality level the organization may have a "hidden factory" of rework that generates high costs. The great product that sells for $20 might cost $30 to make, and the quality costs associated with the product may remain hidden.

Keep the number of measures small. The "family of measures" concept does not mean that we need to measure everything. A typical family of measures contains four to six variables that balance internal and external perspectives (Thor, 1989). Some experts argue that a measurement system should include six to eight key variables (Brown and Svenson, 1988). However, it is possible to have too much of a good thing. Some experts caution that measuring 20 to 30 variables is as bad as not measuring anything at all. It is thought that if people are given more than a handful of objectives to accomplish, they will ignore most of them and concentrate on the two or three they feel are the most important. Tomasko (1990) observes that the new generation of leaders hired to run the leaner, more competitive organizations of the 1990s will have very demanding workloads that will not allow them to monitor more than just a few of the most critical performance indicators.

Provide feedback to those providing "performance data." Even in the best-managed organizations, there is a level of mistrust between workers, management, and executives. The extensive downsizing of organizations has certainly contributed to these feelings. So whenever management announces something new—for example, Six Sigma—the skeptic might conclude that this represents another attempt to control and manipulate the worker for the benefit of the organization. One way to counter such concerns is to share the performance information with the workers themselves; this helps both workers and management make decisions, identify problems, and clarify relationships. It is important to make sure that the feedback is in a form that clearly shows how the information is being used for the benefit of the employees and overall organization: making decisions, identifying problems, or clarifying relationships.

Adjust "monetary" measures for inflation. Occasionally, important metrics will involve units of money. One example is the company's stock price per share. Another is the expense of scrap produced per month. A service example is the average cost of making an in-home warranty repair to a washing machine. A transactional example is a company's annual cost of express delivery service resulting from data entry errors. If one were to plot these monetary values over a certain time frame, the data could be deceiving unless an allowance were made for inflation.

6.3   THE BALANCED SCORECARD

The concept of a balanced scorecard became popular following a research study published in 1990, and ultimately led to the 1996 publication of the standard business book on the subject, titled *The Balanced Scorecard* (Kaplan and Norton, 1996).

Between concept and book there were a number of *Harvard Business Review* articles describing the balanced scorecard methodology in greater and greater detail (Kaplan and Norton, 1996). The authors define the balanced scorecard as "organized around four distinct performance perspectives—financial, customer, internal, and innovation and learning. The name reflected the balance provided between short- and long-term objectives, between financial and non-financial measures, between lagging and leading indicators, and between external and internal performance perspectives."

One basic tenet of the balanced scorecard approach that has met with resistance is that many metrics are better than just a few. For example, Kaplan and Norton suggest using the balanced scorecard to determine the size of executive bonuses based on 13 measures ranging in relative importance from 2.5% to 18% each. They also cite a real-world example of a bank that employed 20 metrics, and another of an insurance company that employed 21 metrics.

The power behind a Six Sigma business strategy is that it focuses on a few measures that are vitally important to the customer and to the organization's strategic goals. One should then identify process improvement opportunities that have the potential to satisfy customer expectations while significantly improving the bottom line of the organization.

A number of organizations that have embraced Six Sigma methodology as a key strategic element in their business planning have also adopted the balanced scorecard, or something akin to it, for tracking their rate of performance improvement. One of those companies is General Electric (GE). In early 1996 Jack Welch, CEO of GE, announced to his top 500 managers his plans and aspirations regarding a new business initiative known as Six Sigma (Slater, 2000). When the program began, GE selected five criteria to measure progress toward an aggressive Six Sigma goal. Table 6.1 compares the GE criteria with the four traditional balanced scorecard criteria. We have ordered the four GE criteria so that they align with the corresponding traditional balanced scorecard measures. The fifth GE criterion, "supplier quality," can be considered either a new category or a second example of the balanced scorecard "financial" criteria. We prefer the latter view.

TABLE 6.1.   Measurement Criteria: GE versus Balanced Scorecard

General Electric	Balanced Scorecard
• Cost of poor quality (COPQ)	• Financial
• Customer satisfaction	• Customer
• Internal performance	• Internal
• Design for manufacturability (DFM)	• Innovation and learning
• Supplier quality	

In today's business climate, the term "balanced scorecard" can refer strictly to the categories originally defined by Kaplan and Norton (1996), or it can refer to the more general "family of measures" approach involving other categories. General Electric, for example, uses the balanced scorecard approach but deviates from the four prescribed categories of the balanced scorecard when it is appropriate. Godfrey (December, 1999) makes no demands on the balanced scorecard categories other than that they track goals that support the organization's strategic plan.

PART 3

SIX SIGMA BUSINESS STRATEGY

7

DEPLOYMENT ALTERNATIVES

Most of the Six Sigma statistical tools and methodologies are not new. From an organization's point of view, the traditional approach to the deployment of statistical tools has not been very effective. Most complex statistical analysis has been left to an internal or external statistical consultant or has been part of a segregated quality department.

An engineer may approach a statistical consultant with no real understanding of how he can benefit from the wise use of statistical techniques. The statistician should first learn the technical aspects of the dilemma presented by the engineer in order to give the best possible assistance. However, most statisticians do not have the time, background, or desire to understand all the engineering dilemmas within their corporate structure. To be effective, therefore, engineers need to have some basic knowledge of statistical concepts so that they can first identify an application of these concepts and then solicit help, if needed.

Easy-to-use computer software has made the process of statistical analysis readily available to a larger group of people. However, if the practitioner does not have a sound understanding of the assumptions and limitations of a statistical methodology, computer-derived models can be erroneous or misinterpreted. Also, even when an engineer has recourse to such software, the issue of problem definition and the application of statistical techniques still exists.

When the Six Sigma business strategy is followed and combined with the wise use of metrics and statistical tools, the strategy can result in dramatic improvements to an organization's bottom line. In this chapter and the others in

Part 3, we describe a successful S⁴ business strategy, including the creation of a Six Sigma infrastructure, training/implementation, and effective project selection. This chapter deals specifically with the deployment portion of the overall business strategy. It discusses the executive's view of the deployment of Six Sigma, using projects as examples and pointing out advantages of this strategy. A sample plan for deploying the S⁴ business strategy within an organization is also included.

To have true success with Six Sigma, executives must have a need, a vision, and a plan. This chapter can help organizations create a plan.

7.1 DEPLOYMENT OF SIX SIGMA: PRIORITIZED PROJECTS WITH BOTTOM-LINE BENEFITS

Often organizations do not look at their problems as the result of current process conditions. However, if they did, their problems could often be described in terms similar to those in Figure 7.1. They could determine key process output variables (KPOVs) of their process such as a critical dimension, overall cycle time, DPMO rate, and customer satisfaction. By sampling this output over time at a frequency that gives them a "30,000-foot level" view, they could gain insight into how KPOVs are changing, detect trends in the process, and distinguish between common and special cause variation. This data could then be transformed and combined with data on customer needs to get a picture of the accuracy and precision of the overall process.

Instead of looking at their situation as a process, organizations often react over time to the up-and-down movements of their products, reacting to KPOV levels in a "firefighting" mode and "fixing" the problems of the day. They often have multiple "process improvement" projects working simultaneously. Not realizing that "noise" (material differences, operator-to-operator differences, machine-to-machine differences, and measurement imprecision) can impact a KPOV to a degree that results in a large nonconforming proportion, organizations may make frequent, arbitrary tweaks to controllable process variables. Practitioners and management might even think that this type of activity is helping to improve the system. However, in reality they may be spending a lot of resources without making any process improvements. Unless holistic process changes are made, the proportion of noncompliance, as shown in Figure 7.1, will remain approximately the same.

Organizations that frequently encounter this type of situation have much to gain from implementing the S⁴ business strategy. Often, when organizations assess their Six Sigma needs, they do not give adequate attention to the costs of not making a change. The concept of "the cost of doing nothing," described earlier in this book, addresses these costs and leads to a better appreciation of the potential gain from the implementation of Six Sigma.

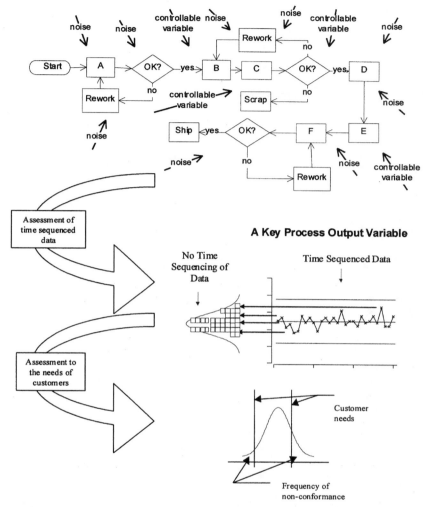

Figure 7.1   Example of a process with a key process output variable [Reproduced from Breyfogle (1999), with permission]

As illustrated in Figure 7.2, the strategy after assessment and kickoff consists of a series of phases:

- Deployment phase, which includes the Define Phase for projects: determining what is important to the customer
- Measure Phase: understanding the process

- Analysis Phase: determining the major causes of defects
- Improve Phase: removing the major causes of defects
- Control Phase: maintaining the improvements

These phases are referred to as the DMAIC methodology. Details of what is to be accomplished and associated challenges of each phase are illustrated via examples included in later chapters.

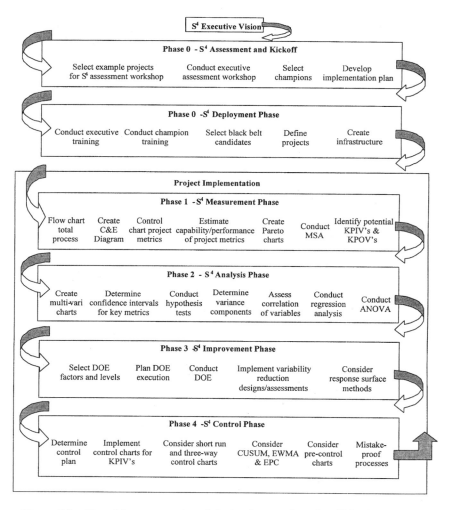

Figure 7.2   Pictorial representation of the implementation of an S^4 business strategy [Reproduced from Breyfogle (1999), with permission]

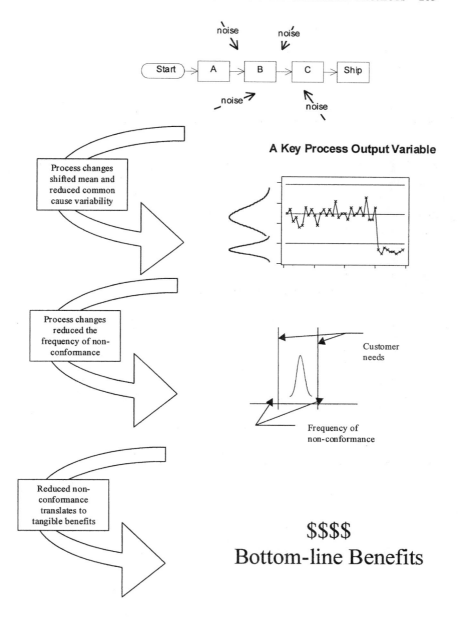

Figure 7.3   Example of process improvement and impact to a key process output variable [Reproduced from Breyfogle (1999), with permission]

When a Black Belt uses the steps summarized in Figure 7.2 as a project during or after training, the type of process improvement exemplified in Figure 7.3 can result. This can occur because processes were simplified or products were designed that required less testing and became more robust or indifferent to the noise variables of the process. This effort can result in an improvement shift of the mean and a reduced variability that leads to quantifiable bottom-line monetary benefits.

7.2   ADVANTAGES OF DEPLOYING SIX SIGMA THROUGH PROJECTS

Some Six Sigma training organizations have created a university-type learning environment that focuses on teaching and application of individual tools. A prospective Black Belt candidate within this university environment might be taught, for example, the mechanics and theory of completing a Gage R&R. The student might then need to apply the tool to sample applications. Upon completion of formal training, the student may need to serve an "apprenticeship" under the guidance of a person experienced in applying Six Sigma techniques. When the student has sufficiently demonstrated the application of Six Sigma tools over a length of time, she might then receive some form of certification.

Implementing Six Sigma, as shown in Figure 7.4, through a mere teaching of the tools could yield application successes; however, in this situation there is typically a lack of visibility of the benefits of Six Sigma to upper management, and significant accomplishments may go unrecognized. Any "wins through Six

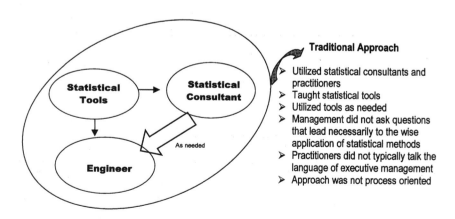

Figure 7.4   Traditional approach to deploying Six Sigma

Sigma"—which themselves can lead to future successful applications of Six Sigma—may fail to make an impact on the organizational culture. In addition, Six Sigma practitioners may have to continue to fight for funds, or face being done away with if the business fails to prosper as expected.

Some organizations offer Six Sigma training that requires the trainees to apply techniques to projects that are assigned beforehand. Instead of focusing on individual tools, this approach teaches a methodology that leads to selecting the right tool, at the right time, for a project. An illustration of our suggested approach for the project integration of tools was given in Figure 7.2. Figure 7.5 summarizes our preferred approach to deploying Six Sigma. Consider the following Six Sigma deployment benefits of using projects instead of individual tools:

- Offers bigger impact through projects tied with bottom-line results
- Utilizes tools in a more focused and productive way
- Provides a process/strategy for project management that can be studied and improved
- Increases communication between management and practitioners via project presentations
- Facilitates a detailed understanding of critical business processes
- Gives employees and management views of how statistical tools can be of significant value to organizations

Deploying Six Sigma through projects instead of through tools is the most effective way for an organization to benefit from its investment in Six Sigma.

7.3 CHOOSING A SIX SIGMA PROVIDER

When an organization chooses to implement Six Sigma, it is very important that a group outside the company be enlisted for help with the implementation. This may initially appear to be an expensive proposition as compared to an in-house implementation. However, the organization has to consider the real cost and time required for effective/interesting course material development, trainer development, and so forth. In addition, it is important for organizations to consider the value of time lost when instructional material is not satisfactory, is not effective, and/or is inefficiently presented.

A good Six Sigma provider can help with setting up a deployment strategy, conducting initial training, and providing project coaching. The decision of which group is chosen can dramatically affect the success of a program. However, choosing the best group to help an organization implement Six Sigma can be a challenge. Often the sales pitch given by a Six Sigma provider sounds good;

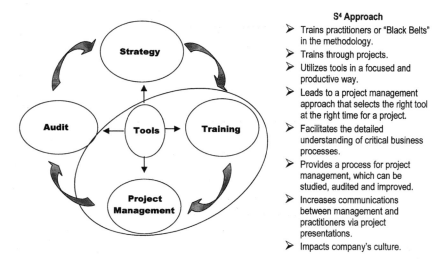

S⁴ Approach

➤ Trains practitioners or "Black Belts" in the methodology.
➤ Trains through projects.
➤ Utilizes tools in a focused and productive way.
➤ Leads to a project management approach that selects the right tool at the right time for a project.
➤ Facilitates the detailed understanding of critical business processes.
➤ Provides a process for project management, which can be studied, audited and improved.
➤ Increases communications between management and practitioners via project presentations.
➤ Impacts company's culture.

Figure 7.5   S⁴ approach to deploying Six Sigma

however, the strategy and/or training material do not match the needs of the organization. When deciding which Six Sigma provider to partner with, we think that it is essential that decision makers view one day of actual Black Belt training from each provider under consideration. During this time these decision makers can also talk with current Black Belt candidates to get firsthand experience about the Six Sigma provider.

We also suggest that the following questions be presented to Six Sigma providers who are being considered:

- What is your basic Six Sigma implementation strategy and flow?
- What do you suggest doing next if we would like to compare the Six Sigma program offered by your organization with that offered by other organizations?
- What book do you use that follows the material taught within the course (so that people can get further information or review a concept at a later date)?
- During Six Sigma training do you use multimedia equipment with a presentation program such as PowerPoint, or do you use transparencies only?
- What is the basic format of your Six Sigma course for executive training and champion training?
- How do you address business and service processes?

- What topics do you cover within your Six Sigma workshops?
- What is the format of the handout material?
- Is licensing of instructional material available for use by our internal instructors?
- Is the material in a form that others can easily teach from?
- Have all the courses that your organization has sponsored used your material (some providers subcontract much of their workshop instruction to other organization that use their own material)?
- Describe the frequency of exercises (i.e., manual, computer software, and other hands-on exercises) within the Six Sigma workshop.
- How do you address application of the techniques to real-world situations?
- Are attendees required to have computers?
- What software do you use? (We suggest using a general-purpose package such as Minitab.)
- Is the software easy to use and learn?
- What have others said about your training/consulting in the application of Six Sigma? Can they be contacted?
- What is the experience level of the person(s) responsible for developing the courseware?
- What companies have you helped successfully implement Six Sigma?

Once a list of providers is narrowed down, consider requesting that they describe their basic implementation strategy to prioritize projects within organizations. Again, we also recommend visiting each prospect to see firsthand a one-day Six Sigma training session in progress. Send a decision maker to view this session. Some Six Sigma providers might initially look good, then appear less desirable after their training approach and material are reviewed firsthand.

7.4   ESSENTIAL ELEMENTS OF AN S⁴ IMPLEMENTATION PLAN

Applying Six Sigma so that bottom-line benefits are significant and lasting change results requires a well-thought-out and detailed plan. Once a Six Sigma provider is selected, it should assist the organization with the development of an effective implementation plan. Figure 7.6 shows a sample implementation schedule for the kickoff of a Six Sigma plan. Although every plan is unique, it should consider the following essential elements, which are described in detail in subsequent chapters:

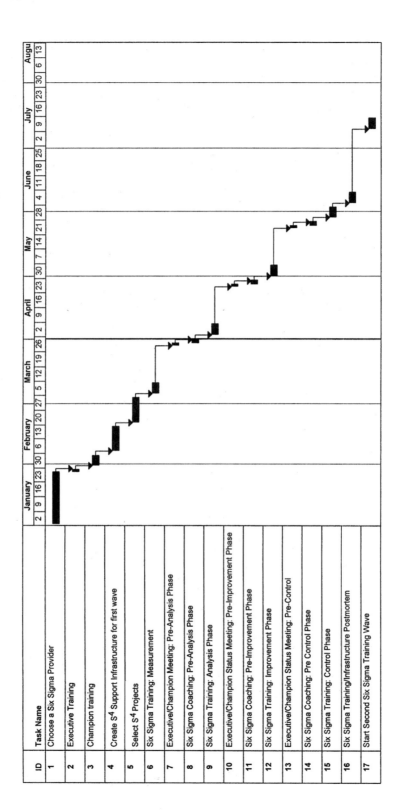

Figure 7.6 Kickoff schedule for Six Sigma

Create S^4 Support Infrastructure for First Wave

For Six Sigma implementation to be successful, there must be commitment from upper-level management and an infrastructure that supports this commitment. The key factors in creating a successful Six Sigma infrastructure within an organization are discussed in detail in the next chapter.

Selecting Key Players

Deployment of S^4 techniques is most effective through Six Sigma practitioners—often called Six Sigma Black Belts—who work full time on the implementation of the techniques through S^4 projects selected on business needs. To have high-level managers who champion these projects, direct support needs to be given by an executive management committee. Organizationally, assigned individuals—often called Six Sigma Champions—orchestrate the work of Six Sigma Black Belts in the strategic business unit for which they are responsible. In matters of technology, Six Sigma Black Belts can be supported by others—often called Six Sigma Master Black Belts—who are considered experts at the implementation of Six Sigma methodology and associated statistical tools.

Selecting Key Projects

Initial projects need to focus on the transfer of customer information to improvement teams. Teams will be successful only if they focus on bridging the gap between what their key process produces and what customers really need. The most successful projects have hard financial targets and meet customer CTQs simultaneously. Also, S^4 projects should be of manageable size with respect to the impact of the overall system improvements.

Training and Coaching

Prior to training registration, trainees should have been assigned to a defined project that they will be working on. This will help them to understand how to specifically utilize tools as they apply to the process they will be analyzing. It will also give them an opportunity to ask detailed questions and receive coaching on their projects.

There is not a one-size-fits-all training approach for the different types of key players. Effective training for the various roles of Six Sigma will vary in content and duration, as described later in this section.

Project Report-outs

Although difficult to schedule, ongoing top-down and down-up communications are also essential elements of a deployment plan. Communication plans are often the first thing to get overlooked on a busy manager's schedule but are essential to break down barriers between management and practitioners.

It is important that executives ask the right questions that lead to the right activity, sometimes referred to as the "five whys." This is an informal tool that management can use to help practitioners get at the root cause of the defect. By continuing to ask "why" after the team develops solutions for the problem, one can help practitioners identify areas that need more focus or analysis. It will not always take five iterations of "why does this defect occur" to get at the root cause. Leaders must take care when using this approach so that it comes across as a mentoring tool instead of a tool used to judge the effectiveness of the team. Questions should always be focused on understanding the process better, not judging the team's analysis of the process.

The quality community has complained for years that management has not listened and supported their quality initiatives. To gain the awareness of upper-level management to the value of statistical analysis, effective project reports are needed that are scheduled up front. Quality professionals often need to do a better job communicating to executives in clear language. Project report-outs should tell a simple story, using easy-to-understand graphs that highlight the root cause of the problem.

Project report-outs should also be structured up front. Management needs to define the frequency and expected content of these reports. Project report-outs serve the purpose of supplying periodic feedback on the status of the project, which will aid in its timely execution. Initially, consider report-outs after each phase in order for management and the attendees to learn the methodology together. Later, one might consider streamlining the process. General Electric went from reporting out after each phase to reporting out before any project improvements were made and upon completion of the project. Also, as part of an overall communication plan, results on successful projects should be posted for all to see; this will serve to generate enthusiasm for Six Sigma by showing its possibilities.

Postmortem

Successful deployment of Six Sigma is best achieved in a series of waves, each wave focusing on strategic change areas. Between waves, there is time for evaluating effectiveness, compiling lessons learned, and integrating improvements into the infrastructure. When a Six Sigma infrastructure is created in waves, one can take time between waves to assess progress and incorporate critical lessons learned for lasting change.

At General Electric, when transitioning from wave 1 into wave 2 during the implementation of Six Sigma, "postmortem" discussions were conducted. Subsequent brainstorming and planning sessions led to the integration of the following valuable lessons learned in the deployment strategy:

- Master Black Belts were aligned with strategic business units (SBUs) and worked together to leverage projects and create strategic focus areas.
- The number of Black Belts and Green Belts was significantly increased.
- Mentoring relationships were established between Black Belts from wave 1 and new Green Belts/Black Belts.
- The Define Phase was added to the Measure-Analyze-Improve-Control deployment training to assist the team in scoping manageable-sized projects.
- Project report-outs were more structured and less frequent.
- Larger-theme projects were developed that focused on key areas, with numerous structured projects relating to the theme being addressed simultaneously.
- An annual Six Sigma week was held that allowed quality practitioners to share the past year's successes, communicate lessons learned, and elect the project of the year.
- Training was made mandatory for promotion.

8

CREATING A SUCCESSFUL SIX SIGMA INFRASTRUCTURE

Six Sigma can be a great success or a great failure, depending on how it is implemented through the corporate infrastructure. Creating a successful Six Sigma infrastructure is an ongoing process whose aim is to infuse an awareness of quality into the way all employees approach their everyday work. Infrastructures can vary significantly, depending on the culture and strategic business goals of organizations. Every organization's creation of a Six Sigma infrastructure is unique; however, there are factors common to every success story.

Figure 8.1 shows an example of an interrelationship digraph (ID) of Six Sigma success factors and how they interact. This ID captures the experience of one of the authors (Becki Meadows) with GE's worldwide implementation of Six Sigma in 1995. A high number of outgoing arrows in the diagram indicates that the item is a root cause or key driver of the success of Six Sigma and should be addressed initially. A high number of incoming arrows indicates a key outcome item that will be realized if key drivers are addressed. A summary ID shows the total number of outgoing and incoming arrows in each box.

This ID indicates how executive involvement was a key driver in the realization of the key outcome: delivering real results. Key drivers in descending order of importance are:

- Executive leadership
- Customer focus
- Strategic goals

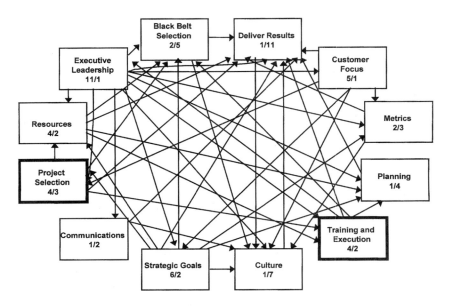

Figure 8.1 An interrelationship digraph assessing Six Sigma success factors and their interrelationship

- Project selection
- Training and execution
- Resources
- Black Belt selection
- Metrics and feedback
- Culture
- Communications
- Planning
- Results

An ID for any one company may produce an order different from above; however, many of the factors will be like those listed. This chapter discusses the critical success factors depicted in Figure 8.1, excluding the factors outlined in bold, "Project Selection" and "Training and Execution," which are covered in other chapters. This chapter also includes a sample implementation schedule and lessons learned from other companies that have created successful Six Sigma infrastructures.

8.1 EXECUTIVE LEADERSHIP

Executive leadership is the foundation of any successful Six Sigma implementation. Most companies that achieve significant results with Six Sigma have the commitment and leadership of their executive management. As Figure 8.1 illustrates, without executive involvement, any other factors critical to the success of Six Sigma will not be as effective. General Electric's tremendous success with Six Sigma is compelling evidence of this assertion.

When GE employees initially referred to Six Sigma as just another "flavor of the month," Jack Welch, GE chairman and CEO, changed the business structure at the corporate level to underscore the importance of the goal (Miles, 1999). Welch made Six Sigma training mandatory for any employee who wanted to be considered for promotion, including senior executives, and forcefully reiterated GE's mission of becoming a Six Sigma company within five years.

Implementation needs to originate at the top echelons of an organization. Key stakeholders have to be identified and committed up front. To kick off Six Sigma, consider holding a series of executive retreats. Goals for these meetings could include:

- Generating executive-level enthusiasm for Six Sigma
- Establishing a team of Six Sigma leaders with specific responsibilities
- Understanding the overall Six Sigma methodology
- Defining how Six Sigma complements current business strategies
- Brainstorming communication plans
- Allocating resources and naming key players
- Creating a Six Sigma road map/implementation plan

For brainstorming sessions, consider using a trained facilitator to help create an ID similar to Figure 8.1. This is an excellent aid in defining the specific factors critical to the success of Six Sigma. After key focus areas are identified, more brainstorming sessions can be conducted to determine action plans for addressing the key drivers.

Whether Six Sigma succeeds or fails depends to a large extent on how well executives understand the value of the S^4 methodology and sincerely promote it within the company. An executive retreat can help identify true champions who will promote change and make employees aware of quality in their approach to everyday work. It can also help prioritize principles and actions necessary in establishing an implementation road map. Through discussion and planning, employees can orient themselves in such a way that their individual experience with execution of Six Sigma is successful.

8.2   CUSTOMER FOCUS

Focusing on the needs of customers goes hand in hand with creating a successful Six Sigma infrastructure. The factors that are important to customers are a necessary input for the process improvement team's true success. Therefore, evaluating the customer's perception of quality should be at the forefront of an implementation process and a foundation block of an infrastructure.

Complaints from customers should be viewed as an opportunity for growth and increased market share. These complaints should serve as a spotlight on areas needing process improvement. The key to success in this initial step is to make it easy for customer input to be heard. There are various methods to do this, including the following:

- Walking the customer process
- Performing customer surveys
- Conducting personal interviews with key customers
- Establishing feedback/complaint systems
- Developing customer panels

Customer surveys can be an easy and effective means of inserting the voice of the customer into the project-selection process (an employee opinion survey was described in Example 5.3). When General Electric initiated its Six Sigma strategy, each division performed intense customer service surveys. Key customers were asked to rate GE on a scale of 1 to 5 on several critical-to-quality issues. GE also asked its customers to rate the best in class in the same categories, measuring a defect as anything less than 4, and tracking customer satisfaction on a quarterly basis. Areas with low scores were viewed as potential Six Sigma projects.

The following are general process steps for conducting effective surveys within an organization (Folkman, 1998):

1. Hold a steering committee meeting
2. Conduct a focus group and individual interviews
3. Review interview issues, and select strategic issues to build the survey
4. Customize the survey
5. Administer the survey
6. Digitize the survey
7. Analyze reports to identify key issues
8. Present feedback to top management
9. Conduct feedback sessions with managers
10. Hold feedback sessions with those who took the survey

11. Create action plans
12. Monitor progress
13. Evaluate and make modifications to the survey

In step four, customize the survey, care must be taken to create a survey that will yield valuable feedback and lead to actions that make a difference. The following guidelines will increase the reliability and validity of a survey (Folkman, 1998):

- Each survey item should be a concise statement using simple language (not a question).
- Items should be stated mildly positive or mildly negative.
- Items should measure only one issue at a time.
- Once the item is constructed, determine its effectiveness by considering what the feedback would indicate and what corrective actions would result from very positive or very negative feedback.
- Test the survey on a few people to sample the quality of the feedback.
- Include a few open-ended questions, which will provide valuable feedback even though they are difficult to analyze.
- Use a five-point Likert Scale.

When analyzing customer surveys, it is important to have an overall plan to increase efficiency and draw a simple story from the data (Folkman, 1998). This plan may encompass the following items:

- Return rates
- Overall scores
- Survey categories
- Strengths and weaknesses (high and low scores on individual items)
- Correlation analysis (which individual items correlate strongly/weakly with overall scores)
- Demographics
- Written themes

Care must be taken that the data are statistically significant and do not represent special circumstances, such as only one customer point of view.

At Smarter Solutions, we incorporate the voice of our customers into our business strategies by creating feedback loops on our critical processes. Table 8.1 contains a survey we created to gain insight on a process whose output you are very familiar with—the writing of this book. This is only a sample survey, but we would appreciate your candid feedback. We hope you will take the time to copy it, fill it out, and return it to us. Or you can visit our website at www.smartersolutions.com and fill out the survey on-line. We will use your feedback as we improve on this book in subsequent editions.

TABLE 8.1   Customer Satisfaction Survey of This Book

Please mark an X in the column that most closely matches your view for each of the statements below:	Strongly Disagree	Disagree	Neutral	Agree	Strongly Agree
1. Book organization was logical and cohesive.	❏	❏	❏	❏	❏
2. Glossary at the end of the book was helpful.	❏	❏	❏	❏	❏
3. My organization will use the book as a guide to implementing Six Sigma.	❏	❏	❏	❏	❏
4. "Frequently Asked Questions" section was useful.	❏	❏	❏	❏	❏
5. Explanation of Six Sigma in statistical terms was understandable.	❏	❏	❏	❏	❏
6. Satisfied with the purchase of this book.	❏	❏	❏	❏	❏
7. Development applications of Six Sigma were explained well.	❏	❏	❏	❏	❏
8. Cross-references to Breyfogle (1999), *Implementing Six Sigma,* were useful.	❏	❏	❏	❏	❏
9. Length of the book was appropriate for managers and executives.	❏	❏	❏	❏	❏
10. Written text was concise and informative.	❏	❏	❏	❏	❏
11. Service/Transactional applications of Six Sigma were explained well.	❏	❏	❏	❏	❏
12. Education and training requirements of Six Sigma were adequately covered.	❏	❏	❏	❏	❏
13. Manufacturing applications of Six Sigma were explained well.	❏	❏	❏	❏	❏
14. Relationship of Six Sigma with other quality systems was insightful.	❏	❏	❏	❏	❏
15. Index at the end of the book made it easy to find key Six Sigma topics.	❏	❏	❏	❏	❏
16. Figures and tables were clear and useful in explaining text information.	❏	❏	❏	❏	❏
17. Key Six Sigma roles for people were clearly defined and differentiated.	❏	❏	❏	❏	❏
18. Flow of topics within the book was good.	❏	❏	❏	❏	❏
19. Infrastructure requirements for a Six Sigma strategy were well articulated.	❏	❏	❏	❏	❏
20. The project integration of the tools and examples in the application sections were explained well and serve as good references to project execution.	❏	❏	❏	❏	❏

Comments/Improvement Recommendations:

Occupation:

Please send completed copy to: Smarter Solutions, Inc., 1779 Wells Branch Parkway, #110B-281, Austin, TX 78728. Or complete an online version at www.smartersolutions.com.

Learning through customer feedback what works and what doesn't will establish the mind-set of continual process improvement within an organization. Jack Welch himself has been quoted as saying that a business strategy alone will not generate higher quality throughout an organization. Depending on the size of an organization and its core values, the word *customer* can take on many different definitions. When collecting feedback, care should be taken to maintain a comprehensive view of customers. By combining external feedback with such things as internal business strategies, employee needs, and government regulations, organizations can obtain a balanced list of customer needs.

8.3 STRATEGIC GOALS

Once the needs of key customers are collected and analyzed, a method is required to transform them into strategic goals for the organization. At this stage in the implementation process, quality function deployment (QFD), or the "House of Quality," is a useful tool to drill down from customer needs to strategic focus areas that have a bigger impact on the bottom line of a company.

Figure 8.2 summarizes the customer satisfaction House of Quality process. The overall approach is an iterative process in which a cross-functional team completes a series of "houses" via the following high-level steps or guidelines:

1. *"Primary What"*: Insert a comprehensive list of balanced customer expectations in the first column of "House 1" with the importance ratings from customer surveys inserted in the adjacent column. Again, it is a good idea to insert strategic business objectives here for a balanced list of "primary whats" along with any government regulations or employee needs.

2. *"How"*: A brainstorming session is held with a cross-functional team in the second substep to determine the important "hows" relating to the initial list of "whats." These "hows" form the primary row of "House 1." They are an organization's high-level business processes and address how the customer requirements can be met.

3. *"Relationship Matrix"*: Relationships of high, medium, low, and none are assigned to how business processes affect customer requirements. Numerical weights such as 9, 3, 1, 0 can be used to describe these relationships. Calculations are then performed: cross-multiplication with the customer important rankings. The values within columns are then totaled.

4. *"How Much"*: The output is a ranked list of initial "hows" that can be used for strategic focus areas. These items are then transferred to "House 2," becoming the next "whats" if more detail is needed.

5. The process continues until the ranked "hows" are specific enough to assign strategic goals and measurements. The subsequent further scoping of the final "hows" into S⁴ Improvement projects is described in Chapter 10 of this book, under "Project Selection."

Upon the successful completion of this process, there will be a list of key focus areas, which can be assigned strategic goals and objective measures that serve as S⁴ project opportunities. This process provides confidence that project resources are focused on meeting critical needs of the business and the needs of customers.

The following are tips to consider when creating a customer satisfaction House of Quality:

- The more specific the outputs from customer surveys, the less time the QFD process will take.
- Combine customer input with internal business strategies, employee needs, and government regulations to obtain a comprehensive list of customer needs.
- Consider first organizing all the "whats" into the categories with similar detail, and then perform the iterative process, inserting new "whats" at the appropriate "house" level.
- Bias is easily injected into the process through the wording of survey questions or the choice of team members.
- The process can require a lot of time and resources to conduct.
- Hold separate meetings for the brainstorming sessions so the output is meaningful and participants don't feel rushed.
- Sometimes it takes many repetitions of the process to achieve meaningful results.

The customer satisfaction House of Quality process is time consuming and is often skipped over, but if done with care, it is a valuable investment in the lasting success of projects. It is also a valuable step in effective resource planning, helping to prioritize project work to meet resource constraints. Once created, it can serve as a backlog of potential S⁴ projects, so that once a Six Sigma practitioner is finished with a project, she can easily move on to another without starting the project selection process over and over again. Accurate QFD investigations can be useless if they are not conducted and compiled in a timely fashion. Some situations do not necessarily require all the outputs from a formal QFD. Chapter 10 addresses other project considerations when a list of strategic projects already exist.

For further detail on fully utilizing all aspects of QFD, please refer to Chapter 13 of Breyfogle (1999), *Implementing Six Sigma*.

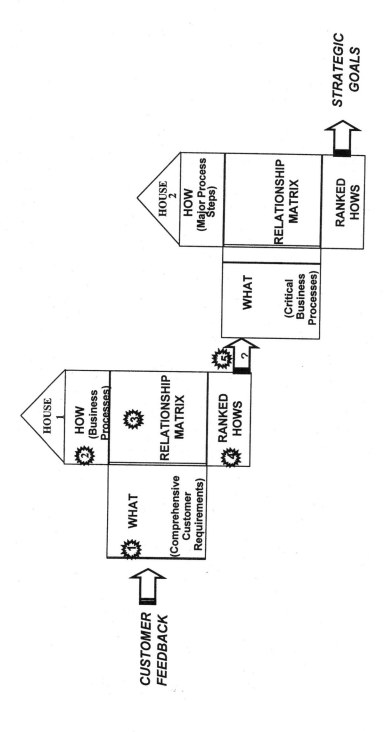

Figure 8.2 Customer satisfaction House of Quality: Relating customer needs to critical business processes

123

8.4 RESOURCES

The methodology of Six Sigma needs to be combined with the right people for real results (Harry, 2000). GE has had much success with the deployment of Six Sigma techniques through individuals called Black Belts, who work full-time on executing Six Sigma projects (Slater, 1999). These individuals are the project leaders, but they cannot do the work of Six Sigma alone.

For process teams to achieve results that are meaningful, the priorities of team members need to be aligned to the strategic goals of the projects. Management needs to create a supportive environment and realign resources to priorities when needed. Listed below are the roles and responsibilities of a well-staffed team as it existed within GE. Smaller companies may need to combine roles if resources are limited. Also, larger project themes may require additional resources.

Champion (Executive-Level Manager)

- Remove barriers to success
- Develop incentive programs
- Question methodology
- Approve or reject project improvement recommendations
- Implement change
- Communicate the Six Sigma vision
- Determine project selection criteria
- Approve completed projects
- Allocate the necessary resources to ensure project success

Master Black Belt (Statistical Expert)

- Complete extensive Six Sigma training
- Coach multiple Black Belts
- Communicate the Six Sigma vision
- Leverage projects and resources
- Share Six Sigma methodology expertise
- Function as a change agent to leverage new ideas and best practices
- Improve overall project execution efficiency
- Conduct and oversee Six Sigma training
- Formulate business strategies with senior management
- Aid in selecting projects that fit strategic business needs

- Facilitate lessons learned session between Black Belts and other Master Black Belts
- Motivate others toward a common vision
- Coordinate activities to drive project completion
- Participate in multiple projects
- Remove barriers to success
- Aid Black Belts in formulating effective presentations to upper management
- Approve completed projects
- Become certified as a Master Black Belt

Black Belt (Project Manager/Facilitator)

- Lead change
- Complete extensive Six Sigma training
- Communicate the Six Sigma vision
- Lead the team in the effective utilization of the Six Sigma methodology
- Select, teach, and use the most effective tools
- Possess excellent interpersonal and meeting facilitation skills
- Develop and manage a detailed project plan
- Schedule and lead team meetings
- Oversee data collection and analysis
- Establish measurement systems that are reliable (when needed)
- Sustain team motivation and stability
- Communicate the benefit of the project to all associated with the process
- Track and report milestones and tasks
- Calculate project savings
- Interface between finance and information management (IM)
- Monitor critical success factors and prepare risk-abatement plans
- Prepare and present executive-level presentations
- Become a certified Black Belt
- Mentor Green Belts
- Complete four to six projects a year

Team Member

- Contribute process expertise
- Communicate change with other coworkers not on the team

- Collect data
- Accept and complete all assigned action items
- Implement improvements
- Attend and participate in all meetings
- Be motivated

Sponsor/Process Owner (Manager of Process)

- Ensure that process improvements are implemented and sustained
- Communicate process knowledge
- Obtain necessary approval for any process changes
- Communicate the Six Sigma vision
- Select team members
- Maintain team motivation and accountability

In addition to the roles listed above, an organization may choose to appoint Green Belts with the same responsibilities as Black Belts, but on a part-time basis. Some organizations set up a mentor relationship in which Green Belts assist Black Belts. Common requirements for Green Belts are to complete up to two projects per year. Other part-time members could include representatives with a financial background to approve monetary calculations and representatives with information technology experience to assist with data collection. The latter is critical to establishing measurement systems that are easily reproducible and reliable. Also, larger companies may want to consider adding an overall Quality Leader to monitor the Six Sigma business strategy on a broad level.

In Figure 8.3, we describe sample organization charts of how a company may choose to relate the above roles to each other within an organization. The top chart shows how Six Sigma might fit into an organization depending on its size and number of strategic business units (SBUs). To achieve maximum effectiveness, executive-level managers of SBUs should champion projects within organizations. This will minimize the risk of creating an "us versus them" environment of some corporate initiatives. Process owners or sponsors should be selected from managers within the department who have ownership over critical business processes. The Black Belts are then assigned to appropriate project sponsors. This Black Belt/sponsor relationship does not have to be one-to-one. Depending on their functional expertise and strategic business needs, Black Belts may sometimes have multiple projects with different sponsors. Sponsors should be chosen on the basis of their ability to lead change and shouldn't be assigned a project just because of position.

Black Belts are to be supported in the management chain of an organization by way of a champion. Technically, they receive support from a Master Black Belt. The bottom chart in Figure 8.3 shows an example of the technical rela-

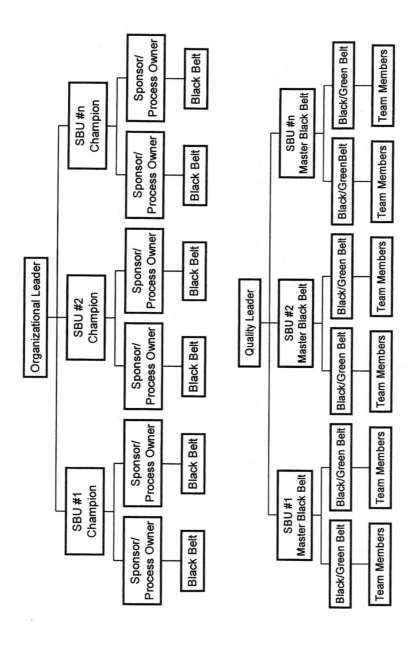

Figure 8.3 Organization charts for six sigma implementation

tionships of an effective Six Sigma organization within GE. In this model, the company appointed an overall quality leader to drive program effectiveness. Master Black Belts report to the quality leader and are responsible for multiple Black Belts or Green Belts within specific departments or functional areas.

Champions, Sponsors, Master Black Belts, and Black Belts are the leaders of Six Sigma change, and care should be taken to select the right players. However, having dedicated team members is just as important as a well-trained leader. Free up time and resources for improvement projects—team members should not feel overburdened by their Six Sigma responsibilities. Black Belts need a supportive environment that prioritizes their projects. Up-front work needs to be done within the team to communicate expectations and establish how to meet them. For additional information, refer to the section "Create Project Charter and Overall Plan" in Chapter 10.

Six Sigma Black Belt Selection

Six Sigma Black Belts need to have many qualities to be effective. They need strong leadership and project management skills. They also need to be aware of decision alternatives relative to choosing the right Six Sigma tool for a given situation.

With respect to technical matters, Six Sigma Black Belts need to be aware that pictures can be deceiving because of sample-size issues, sampling techniques, and so forth. They have to be able to conduct significance tests that assess the likelihood of these differences. Six Sigma Black Belts also need to be able to determine if there are alternative approaches to get around issues involving small sample size. For example, they might need to change attribute data collection procedures to a continuous measurement technique. Or they might have to use a fractional factorial DOE to give greater insight as to what should be done, rather than select and test a large sample size collected over a short period of time.

Not only will Black Belts need to analyze information and wisely use statistical techniques to get results, but they will also need to be able to work with others by mentoring, teaching, and selling them on how they can benefit from S^4 techniques. The difference between a Six Sigma Black Belt who produces average results and one who excels is usually his ability to effectively facilitate his team. In their book *Managers as Facilitators*, Richard Weaver and John Farrell (1999) define facilitation as "a process through which a person helps others complete their work and improve the way they work together." Effective facilitators exhibit the following characteristics:

- Value collaboration
- Value helping others build on good relationships to get their work done
- Value being a supporter

- Behave in a helping manner
- Constantly monitor what is going on
- Think before they speak
- View themselves as instruments to help the group
- Clearly define what is expected of the team
- Help groups relate to conflict as normal and productive
- Listen actively and effectively to the team

People selected to become Black Belts need to be respected for their ability to lead change, effectively analyze a problem, facilitate a team, and successfully manage a project.

8.5 METRICS

This section describes metrics as they relate to the overall implementation of a Six Sigma business strategy. Metrics tracking the effectiveness of the Six Sigma infrastructure can drive behavior and create an atmosphere for objective decision making. A successful Six Sigma infrastructure needs to be thought of as an ongoing process that can be continually improved. Metrics should be created to provide valuable information on the variability and the efficiency of this process. Suppose a metric is created that measures how many projects are completed each quarter. If, at the end of the first quarter, an organization falls short of its goal, one should first challenge the metric by asking "Does this metric accurately measure the success of the current Six Sigma infrastructure?" If, for example, Six Sigma financial savings and quality improvements far exceeded expectations, then we might conclude that the number of projects was not met, because the size of projects was larger than expected.

General Electric implemented a company-wide project-tracking database, and all Black Belts were required to enter detailed information about their projects. This database captured valuable information, such as:

- Team members
- Key process output variables
- Key process input variables
- Estimated completion date by phase
- Actual completion date by phase
- Six Sigma tools utilized in the project
- Open action items
- Defect definition
- Baseline measurements

- Financial calculations
- Actual project savings
- Lessons learned concerning project execution

From this database, quality leaders were able to glean insight into the effectiveness of Six Sigma.

In general, metrics need to capture the right activity and be designed to give immediate feedback. Successful metrics focus on the process rather than the product or individuals. If employees are forced to enter additional information and don't immediately see how this information is utilized, the quality of reported data in all likelihood will drop dramatically.

8.6   CULTURE

The launching of Six Sigma affords an opportunity to assess the current culture of an organization. Consider the following questions:

- How has your company historically dealt with change initiatives?
- Does your company make consistent changes that don't last?
- How effective are your project teams?
- Are you frequently focusing on the same problem?
- How do your employees attack problems when conducting their daily work?
- What is required within your company culture to make continual process improvement a lasting change?
- What will prevent your company from achieving success with Six Sigma?

In today's constantly changing marketplace, companies that are able to embrace change in a focused manner are the leaders in their field. Many companies attempt to improve products with numerous small changes, or "tweaks," to their current processes. However, frequently changes are not documented, nor are the associated results. Substantial results are rarely obtained with this half-hearted method of change. Typically, when employees in this type of corporate culture hear of a new initiative they wonder what will be different.

To move from an existing culture that is skeptical of process change to a culture that embraces continual process improvement requires an understanding of forces for change and forces that promote staying the same (Weaver and Farrell, 1999). A *force field analysis* is an effective way of illustrating the forces in an organization that drive toward a Six Sigma solution and those working against it. Figure 8.4 shows an example of what might be expected from this type of analysis, where the weight of the line is an indication of the importance of a force.

CULTURE THAT EMBRACES SIX SIGMA

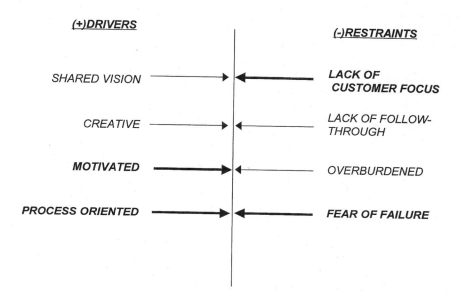

Figure 8.4   Force field analysis

The next step is for a company to create strategic plans to overcome their restraining forces and enhance their drivers. Listed below are examples of company cultures that exhibit some of the key drivers and restraints listed in Figure 8.4:

Creativity

Creative thinking must be at the heart of any process improvement project. Teams need the freedom to experiment, to be supported in playing with the process. Lack of creativity is a lurking problem that is not always evident at the onset of Six Sigma. As companies move from low-hanging-fruit projects, "out-of-the-box thinking" is required to continue process improvement efforts.

3M encourages scientists to take 15% of their time to pursue ideas, projects, and activities of their own choosing (Fagan, 2000). This rule set the creative environment in which Post-it self-stick removable notes were conceived; these notes have been a very successful 3M product. The company brought creative thinking to the foreground and reaped a substantial reward.

Motivation

Most companies that readily adapt to change have a high level of motivation throughout their workforce. Employees are personally motivated for different reasons, but most employees are motivated in a professional context by the prospect of reward. Remember, reward does not necessarily mean money. A simple, personal message of sincere thanks can sometimes be more effective than a check in the mail. What motivates an employee to risk change? The answer may be different for different departments within a corporation, and can vary from time off to verbal recognition or from new equipment to a weekend getaway. However, all good reward systems are fair and honest and have follow-through.

At GE, motivation was a key driver to success with Six Sigma. In 1995, Jack Welch sent a memo to his senior managers stating that every employee must have begun Six Sigma training to be promoted, leading the Six Sigma charge by connecting quality to compensation (Slater, 1999). To reinforce this charge, he told top executives that 40% of their bonuses would be tied to the successful introduction of Six Sigma. The memo launched the new quality initiative with a fierce momentum. A direct connection between outcome and reward was established; in other words, reward was tied to achievement of breakthrough improvements. GE was consistent with rewarding desired outcomes, and it brought the company true success in its Six Sigma initiative. After a small net loss on GE's initial investment, the company's payoff in a short period of time reached more than $750 million in 1998.

Empowered

Powerful results come from synergistic teams. Without common purpose and shared methods on how to work together, the output of the team will merely be a sum of the outputs of the individual contributors.

In his book *Managing Upside Down*, Tom Chappell (1999) describes how his family-run natural personal-care company (Tom's of Maine) achieved rapid growth and profitability by "managing upside down." Mr. Chappell invented the idea of "acorns," which were small, flexible teams of no more than three people responsible for new idea generation, market research, and scientific investigation. The company had previously gone three years without the introduction of a new product; however, once acorns were invented, the company increased its product line from 27 to 117 in two years, proving it could create profitable products by empowering people to think and act creatively.

Lack of Customer Focus

Customer satisfaction can be defined in terms of quality, cost, and on-time delivery. These three characteristics translate to every business process within a corporation. Yet often within business segments of an organization there is no established link between customer values and the process that creates the product or service. Customer complaints are valuable data that can lead an organization to identify the source of process errors.

At one point in time, Disney World received numerous written complaints from parents stating that their children were not able to meet Mickey Mouse (Hiebeler et al., 1999). Disney viewed the letters as an opportunity to improve its existing customer service and made prompt changes to improve the visibility and access to all of their characters, especially Mickey Mouse. The changes not only addressed the complaints of customers who wrote letters, but also made many of its more than 70 million annual visitors extremely satisfied.

Change requires an honest look at the current culture of an organization. It also takes courage and an environment that supports creative risk. Every organization develops differently. Actions that support change and work well for one company may not work well for another organization. Table 8.2 lists examples of common intervention activities associated with driving change within organizations (Weaver and Farrell, 1999).

TABLE 8.2   Example of Action Items That Support Change

Restraining Forces	Action Items
Programs reinforce the old	Create reward and recognition programs that reinforce the new
People do not have skill sets needed to succeed	Create training programs so that people have the opportunity to learn new skills
Phrases still refer to the old	Introduce new vocabulary so people know you are speaking of the new
Current cultural supports staying the same	Change culture so that it supports the new
People are concerned about making mistakes	Reassure them that mistakes are learning opportunities and failure will not be punished
People are concerned about how changes affect their jobs	Provide information that clarifies how performance will be evaluated
People are isolated in the ways they are going about the new	Celebrate successes
People wonder how serious the company is about this change	Provide consistent communications that clarify how serious the company is about completing this change

Frequently, the failure of a Six Sigma project can be attributed to the inability of the company to embrace and sustain change. The only effective way to stimulate change is to link immediate and unbearable pain to our old behaviors while linking immediate and pleasurable sensations to new ones (Robbins, 1991). The challenge of leaders beginning the Six Sigma journey is to create a culture that is able to absorb the change required to continuously improve.

8.7 COMMUNICATIONS

Typically, company leaders implement Six Sigma because they possess a clear vision of what their company can achieve. Frequently, however, they do not realize the power of effectively communicating this vision throughout the corporation. As mentioned in chapter 7, traditional approaches to deploying statistical methods often fail because of communication issues between management and employees. Often, managers do not ask the right questions to strategically focus the application of statistical methods on process improvement instead of firefighting.

To avoid this situation, executives need to get everyone engaged and speaking the language of Six Sigma. A *shared vision* of how Six Sigma fits the strategic needs of the business should be created. A well-thought-out vision has the power to move people and create shared values. Communicating this vision effectively is an ongoing process and is integral to a well-thought-out Six Sigma implementation plan. The Four C's of effective communication plans are as follows:

Concise

Words should be carefully chosen to communicate the specific goals of the Six Sigma business strategy. These words should become the common language for Six Sigma activities.

Consistent

Communication should be sincere and consistent—that is, it should repeat the same message.

Complete

Communication plans should reach all levels of the organization. If necessary, consider the use of different media. A brainstorming session can help to determine the most effective communication techniques for different groups. One should not underestimate the power of communicating one-on-one, especially with team members working on the shop floor.

Creative

A creative method is needed to acquaint employees with the Six Sigma mission.

The first Black Belts trained at GE were asked to define what Six Sigma was and what it meant to the organization. In a facilitated session, each leader re-created a make-believe one-minute elevator ride with Jack Welch. During the ride, Jack asked what Six Sigma was, what it meant to them, and its importance to the company.

At first, many thought the exercise to be a silly waste of time. But once a few Black Belts went through the process, they realized it was difficult to communicate what Six Sigma meant in a meaningful and concise fashion. After spending time in smaller groups and brainstorming definitions, the Black Belts went through the process again. This time they produced clear, well-thought-out answers that were personally meaningful and easily repeated to other team members.

Communication plans should be carefully considered and executed with enthusiasm. If successful, these plans will be the biggest ally in key stakeholder buy-in. On the other side of the coin, practitioners should be able to construct project reports that speak the language of executives and tell a concise, understandable story. Both employees and management need to make sure that they are asking questions that support finding the root causes on every critical project. Chapter 10 of this book, "Project Selection, Sizing, and Other Techniques," discusses this topic further.

8.8   LESSONS LEARNED

The success of Six Sigma depends on the existence of a solid infrastructure and executive leadership that supports the commitment. The roles of the Six Sigma champions will quickly transform into "watchdogs of quality actions" if executive energy is not infused into the Six Sigma infrastructure. It is crucial that project champions express enthusiasm and belief in Six Sigma. Leaders need to take personal responsibility for driving Six Sigma efforts, including participating in and even running Six Sigma projects. True leaders show belief in a vision that then becomes believable to their employees. True leaders show their employees the future and, through their own actions, motivate them to achieve goals.

Six Sigma was continually communicated at GE as "the way we do business" instead of just another initiative (Slater, 1999). Jack Welch's goal was to have the Six Sigma methodology embraced within all daily activities. He knew that for this company-wide initiative to be a success, everyone had to be involved. He pulled out the slogans and banners in full force, then closely monitored the situation and provided meaningful follow-up when necessary. The following are lessons learned from GE's executive leaders responsible for implementing the Six Sigma initiative (Slater, 1999):

- Quality won't happen by itself
- Avoid putting too much trust in banners and slogans
- To be successful, a quality initiative must be launched on a company-wide basis
- Make sure you have the necessary resources to invest in a new quality initiative (payback)
- Spread the word of the new initiative as quickly as possible
- Senior management needs to lead the effort and play an active, visible role
- Provide monetary incentives for quality success
- Reward aggressive team members
- Monitor all aspects of success
- Make sure the customer is taken into account

When assigning strategic goals within Six Sigma, it is important that the goals focus on meeting the comprehensive needs of customers. These goals must also be SMART: Simple, Measurable, Agreed to, Reasonable, and Time based. Also, a few carefully selected projects that meet certain initiatives will bring more respect and ultimately more buy-in than a shotgun approach to project selection. With respect to strategic goals, Robert K. Phelan, Senior Vice President of Quality Programs at GE American Communications, shares the following lesson: "We experienced false starts when we engaged all employees [in Six Sigma practices] quickly. We quickly reviewed every project and made sure they were aligned with our business objectives" (*Aviation*, 1998).

As mentioned earlier, many times a project fails because a team is not empowered and adequately supported. The jobs of Six Sigma leaders are infinitely more difficult if teams do not have the correct balance of structure, expertise, and resources. However, companies can go overboard with appointing Black Belts. Making every employee either a full-time or part-time Black Belt diminishes the power of the program. If everybody is required to lead a project, that leaves few resources for dedicated team members. A rule of thumb is that full-time Six Sigma practitioners within an organization should number no more than 2% of the total number of employees. Most important, people who become successful full-time Six Sigma Black Belts have a "fire in their belly" to take on process improvement and reengineering challenges. Steve Hollans, a Six Sigma applications expert from AlliedSignal, warns, "An early barrier for us was selecting some people on the basis of their availability instead of people who were capable of being change agents" (*Aviation*, 1998).

At the outset, expect results and develop metrics that lead to the "right" activity. Next, determine what should be done to obtain insight into the source of variability within strategic processes. No matter how solid the infrastructure, the program will never get off the ground without employee buy-in. Metrics need to be established with hierarchy throughout the business. If rewards are allotted only to upper management, the culture and belief system associated with quality efforts will be diluted or easily misunderstood.

Establishing a solid infrastructure can help employees navigate unfamiliar territory and provide them with a framework to make successful decisions. When employees have questions, there will be guide marks to lead the way. Change is confusing and generates insecurity in people. If proposed changes are framed within a structured environment, they are easier to execute and endure. In his article "Leading Change: Why Transformation Efforts Fail," John Kotter (1994) lists the following eight common errors companies make when initiating corporate change efforts:

- Not establishing a great enough sense of urgency
- Not creating a powerful enough guiding coalition
- Lacking a vision
- Undercommunicating the vision by a factor of 10
- Not removing obstacles to the new vision
- Not systematically planning for the short-term wins
- Declaring victory too soon
- Not anchoring changes in the corporation's culture

Organizations can focus on only a limited number of things at once. If a blanket approach is used to apply Six Sigma, buy-in can be low and the effectiveness of the program diminished. Effective planning is required that utilizes the "wave" approach discussed earlier. According to James Baker, president of Delevan Gas Turbine Products Division of Coltec Industries, benefits will be realized more quickly if a specific quality-improvement plan is put in place once employees are trained (*Aviation*, 1998).

Lastly, open all available avenues to share lessons learned. Execution issues that are honestly communicated, with no fear of "finger pointing," are lessons learned and serve as opportunities for process improvements. Learning about the process of creating a successful Six Sigma infrastructure can occur on many different levels. Some effective methods are:

- External benchmarking with noncompetitors
- Periodic "best practices" meetings between Six Sigma practitioners
- Collecting lessons learned in a project-tracking database

Go forward with large goals, but also respect the paths taken by other companies, integrating their most critical lessons learned into a continual process improvement strategy.

9

TRAINING AND IMPLEMENTATION

A simple but significant distinction between education and training is that education teaches *why,* training teaches *how.* Companies need to hire people who already know why and then empower them with a new vision of how. That new vision of how is Six Sigma.

As discussed previously in this book, an education and training phase should be centered around projects. The education component of this endeavor is focused on executives and managers, while the training portion is designed for Six Sigma Black Belts. In this chapter we will focus on the training of full-time Six Sigma Black Belts.

In-depth training for full-time Black Belts is typically delivered at the rate of one week of training per month, followed by three weeks of the participant applying what has been learned to a project within his or her area.

Many tools and techniques should be covered in depth during four weeks of instruction spread over the four-month training period. Statistical techniques can include multiple regression, analysis of variance, statistical process control, Gage repeatability and reproducibility (R&R), process capability and performance capability studies, probability, reliability testing, designed experiments, response surface methodology, basic quality tools, variance components analysis, mixture experiments, and so forth (Breyfogle, 1999). Many of these topics are taught in semester-long courses at a college or university, which can make for an intense four-week course.

To address this, we believe that it is not necessary for trainees to learn detailed theoretical "why" aspects of statistical techniques; it is more important to focus on the "how" aspects. Also, we believe in understanding the basic principles behind the tools and the wise application/integration of these tools with other techniques. The labor-intensive portion of "how" to implement the tools is then addressed through the use of statistical software such as Minitab. With this training strategy, much less knowledge of mathematics/statistics is needed as a prerequisite.

A major goal of S^4 training is to explain how statistical and other tools fit into the overall improvement methodology designed to address strategic issues within an organization. With this approach, more people have the opportunity to apply techniques directly within their own area of the business. We should note that a university or college degree does not imply that a person will be a successful Black Belt candidate. We have seen many trainees with no university or college training have great success implementing Six Sigma.

In truth, training is a lifelong commitment to learning. It is most effective when acquired "just in time" on the shop floor, in the accounting department, or in the bullpen of specialists answering technical-support phone lines. Poor delivery of training material can turn even the best material into a boring exercise that fails to impart useful know-how. Conversely, marvelous trainers are worthless if their materials are outdated or are taught at a level beyond the students' ability to understand. The true test of effective training is not an enthusiastic student evaluation, but rather the students' subsequent ability to perform new tasks effectively on the job.

9.1 STRATEGY

In the training of Six Sigma practitioners, the curriculum needs to provide frequent opportunity to apply Six Sigma tools and techniques. Flexibility, dynamics, and power are added to training sessions when each student has a notebook computer with a statistical software package such as Minitab installed, along with a software suite such as Microsoft Office. A very effective teaching technique is to intermingle computer exercises with presentations of the material using software such as PowerPoint. This technique allows computer exercises to utilize examples, exercises, and data sets that are prepared and loaded onto each student computer before the training sessions begin. For this type of training, we use the examples and exercises described within Breyfogle (1999), *Implementing Six Sigma,* as a foundation.

The material should be organized and delivered in such a way that trainees can immediately apply the concepts to the project they have been assigned— that is, during the three weeks before a new round of instruction begins. During this on-the-job application of Six Sigma tools and techniques, it is important for the instructor and other members of the training faculty to provide students with one-on-one coaching on an as-needed basis. This type of coaching is most effective when it is conducted on-site, where the consultant can become familiar with the processes the trainee must deal with.

The tools and techniques being taught should not be foreign to the students. Examples of how and when to use them wisely is extremely valuable. For this reason, instructors should be individuals who have chartered, trained, and led teams to success on numerous process-improvement projects. They should be able to coach trainees on efficient project execution and give advice in resolving unique dilemmas associated with implementing Six Sigma in their respective organizations.

Our experience in S⁴ training has shown us that the most effective format for the delivery of Six Sigma concepts is as follows:

- Introduce and discuss a *new topic* using a computer projector system in conjunction with a presentation-software package. Transparencies are bulky and discourage frequent updates to the training materials. We recommend avoiding them.

- Present an *example problem* on the same topic, and explore one or more solutions to the problem using real-time application of statistical software, where appropriate, to display or analyze the data.

- Follow up with a *student exercise* using another set of data related to the same topic. Individuals or teams of two students working together should complete the analysis during class, using computers whenever applicable. The problem or exercise should be more challenging than the previous exercise. Consider using larger data sets, slightly different conditions, and typical data anomalies. Of course, two-person teams can reduce the time it takes to solve a problem—and in many cases do so without an instructor's assistance. Utilizing a team approach allows students to share different methods to solving the problem. Instructors are then free to facilitate team discussion and give instruction where needed.

- Periodically present a *large-scale exercise* where teams work together on a generic application of the concepts being taught. In our four-month course for Black Belts, teams collect and analyze data from a variety of sources. The exercises become increasingly complex and demanding as students become more familiar with the source and as their knowledge of statistical tools and techniques grows. We believe that team exercises like these are more valuable than computer models in that they offer a better bridge to the problems inherent in real-world projects. Students also realize that experimental error is now partially under their control.

- Periodically invite *class discussion* on how various tools and techniques are applicable to individual projects. In our experience, these discussion periods are highly valued by students.

It should be noted that Six Sigma tools and techniques are applicable to a wide variety of situations. Therefore, the same Six Sigma course can be used for training people involved in manufacturing, product/process development, service, and transactional processes. However, although many of the same tools can be used in different projects, the sequence of their application may vary. This is one reason why some organizations prefer that training sessions be composed of students from the same discipline to facilitate common group discussions.

We believe it is very beneficial for each student to make brief presentations to other class members explaining how they have applied Six Sigma tools and techniques. In our training programs, each student presents, during the one-week-a-month training sessions, an overview of his project and estimates of bottom-line savings to the organization as a result of the project. These presentations, each limited to about 10 minutes, are scheduled for weeks two, three, and four of the Six Sigma training. Students are strongly encouraged to use the instructor's computer/projector system and their own presentation software. The instructor may provide written feedback of presentations to help students present their project results to their internal management.

There are legitimate differences of opinion on the appropriate guidelines to follow in selecting projects for people training to be Six Sigma Black Belts. Each organization should establish its own guidelines. Some issues to consider when determining these guidelines are listed below. More in-depth information on the project selection process is included in the next chapter.

Projects

Create a process for generating potential S^4 projects that includes the person who will be championing the project. For each S^4 project from a list of possible projects an organization might determine (1) potential cost savings and (2) the likelihood of project success. The product of these two numbers (cost savings and likelihood of success) can then be used to rank-order the list of projects. Draman and Chakravorty (2000) have reported that selecting process-improvement projects with the greatest potential for improving system throughput based on the Theory of Constraints, not just the local quality level, had the greatest impact on the bottom line. They found this approach superior to three traditional TQM-based approaches to project selection: (1) processes producing the highest number of scrap units, (2) processes producing the highest monetary volume of scrap, and (3) treating all processes equally.

An S^4 project should be large enough that its successful completion is significant to the organization, but not so large that it cannot be handled within four to six months. We agree with Jonathan Kozol (Pritchett and Pound, 1997) that one should "pick battles big enough to matter, small enough to win." When an ongoing S^4 project appears to be too large or too small based on these guidelines, rescope the project as soon as possible.

An S^4 project is considerably more likely to succeed when there is someone from within the student's organization championing the project and who has responsibility for the process performance of the area that the project impacts. This helps make the project more visible and integratable into the business area. Students should review the project periodically with this person and important members of the process team. Structured project report-outs with a predefined list of attendees is an effective method of achieving this integration.

Upon completion, the S^4 project should be carefully documented and made available internally to other interested employees. Consider creating a standard project report format that can be used by future students. At a minimum, the format should call for an executive overview, a description of the problem, baseline metrics before work begins, and a flow diagram showing the process before and after improvement. Metrics at the end of the improvement process, results of all tools and techniques that contributed to the problem resolution, and estimated S^4 project benefits may also be included. More information on project report-outs is included in the next chapter.

Metrics and Monetary Issues

The selection of Six Sigma metrics and reporting of these metrics is vital to success. Consider using metrics such as defects per million opportunities (DPMO), process capability/performance indices, control charts, and rolled throughput yield with projects to drive the appropriate activity for improvement and process control. However, we suggest the translation of the impact of Six Sigma initiatives to monetary units whenever possible. Typically, the impact of customer-related metrics described as CTQ or KPOV can even be translated into a monetary impact when some underlying assumptions are made.

The advantage of tying project benefits to money is that this unit of measure can become a common metric across projects. This metric is much easier to interpret and less controversial than a metric such as six sigma quality. Monetary units are also easily understood and valued by all levels of management, both as a current metric and as a metric that has applicability after the initial kickoff of Six Sigma.

Consider whether hard, soft, or both types of savings will be tracked—"hard savings" has tangible monetary benefits, while "soft savings," such as cost avoidance, has an indirect impact. Before making this decision, look at examples 5-2 and 5-3 in Breyfogle (1999), *Implementing Six Sigma*. We believe that measurements drive activity, so it is important to track the right metrics. For an S^4

strategy, we can consider multiple metrics where one of the primary measurements considered should be monetary savings. When a strict rule of counting only "hard" money is applied for this consideration, the wrong message can be sent. A true "hard" money advocate would probably not give financial credit to a Six Sigma project that reduced development cycle time by 25%, since the savings would be considered "soft."

Target S⁴ Project Completion Dates and Stretch Goals

Consider having a somewhat flexible amount of time for project completion after the S^4 workshop session. This flexibility may be necessary to allow for variation in project complexity. An extension of one to three months is reasonable, but you may not want to include this in the original project plan. Perhaps students can request such an extension after all other efforts and attempts have failed to produce more timely results.

It is important to set ambitious, yet attainable, stretch goals for projects. These goals should apply not only to individual S^4 projects, but also to all strategic business goals. If goals are too aggressive, teams may cut corners to meet artificial deadlines.

Recognition/Certification

Carefully consider the process requirements for certification. A multidimensional approach should be used in lieu of only monetary savings—for example, demonstrated project savings of $100,000 or more. Also consider requiring the following: (1) a demonstrated proficiency in the wise use of Six Sigma tools and techniques: (2) the ability to interact appropriately with team members in a variety of roles; and (3) proficiency in documenting all aspects of project activity.

Organizations should also consider how participants within an S^4 business strategy will be recognized in nonmonetary terms. The relative emphasis placed on compensation versus recognition should be looked at carefully.

9.2   AGENDAS FOR SIX SIGMA TRAINING

There are a number of unique roles and responsibilities associated with Six Sigma. It is naive to think that Six Sigma can develop within an organization through a massive hiring of people that are "certified" Six Sigma experts. For Six Sigma to be successful, practitioners need a supportive infrastructure where they can build their expertise.

We will start by describing what a typical four-month Six Sigma training session looks like; our model includes one week of in-class training during each of the four monthly periods. This model will serve as a baseline or reference point against which to describe and quantify the requirements for other types of training.

"Full-Time" Six Sigma Black Belt Training

This training will encompass the Six Sigma themes of measure-analyze-improve-control, which are described in Breyfogle (1999), *Implementing Six Sigma.* Below is the recommended ordering of topics for a four-week training session held over a four-month duration:

Week 1 training (1st month):	(S^{4}Measurement):	Chapters 1–14
Week 2 training (2nd month):	(S^{4}Analysis):	Chapters 15–26
Week 3 training (3rd month):	(S^{4}Improvement):	Chapters 27–33
Week 4 training (4th month):	(S^{4}Control):	Chapters 34–43

It is possible to move some material from the first two weeks to the last two weeks, perhaps eliminating some optional topics from the last two weeks as necessary ("reliability" and "pass/fail functional testing" are considered optional). The four weeks of training may also be reduced to three weeks if certain topics are not considered applicable—"design of experiments (DOE)" and "measurement systems analysis" may be deemed irrelevant to students from an organization based on transactional or service processes (keep in mind, however, that DOE techniques and measurement systems analysis are very applicable to certain transactional or service processes). Hahn et al. (1999) provides a list of key lessons learned at GE relative to statistical tools (see Table 9.1). Consider incorporating these lessons into the training curriculum.

TABLE 9.1   GE's "Lessons Learned" on Advanced Statistical Tools

GENERAL ELECTRIC'S LESSONS LEARNED: TOOLS

- Control variability
- Ensure integrity of measurement system (Gage R&R)
- Key role of designed experiments (and preplanning)
- Usefulness of simulation
- Importance of reliability
- Need to use right tool for the right job
- Existing databases are inadequate for the job
- Don't "oversell" statistical tools
- Avoid diversions

Supporting Six Sigma Green Belt Training

Green Belts do not necessarily need training as extensive as that provided to the full-time Black Belts. For individuals who support Black Belts, our preference is to provide two weeks of training for these individuals, a subset of the four-week training material. As part of this training, projects should be completed with the support of Black Belts. For organizations that assign independent Green Belts to substantial projects, the full four-week training is recommended.

Six Sigma Champion Training

We recommend that the executives who will be championing Six Sigma projects receive five days of training, with a focus primarily on the material from week 1 of a four-week Six Sigma training session and some material from weeks 2, 3, and 4. Emphasis should be given to the selection and management of Six Sigma projects and Black Belts, and to their role as removers of roadblocks in the execution of the projects. We believe that this training should be "hands-on" relative to using statistical software such as Minitab and conducting team exercises. Through this training, managers who are championing Six Sigma projects will obtain a much better appreciation of the work to be conducted by the Black Belt. Through this knowledge, the manager who acts as a team champion can give better guidance and become proficient at selecting projects.

Executive Training

We recommend that S^4 executives receive one to two days of training. This training not only describes the basics of Six Sigma but also emphasizes the establishment and management of a Six Sigma infrastructure. Emphasis should be given to the importance of *wisely* applying Six Sigma techniques, where one-on-one dialogue is encouraged to address specific issues and implementation questions.

Master Black Belt Training

Six Sigma Master Black Belts are trained Six Sigma practitioners who have become full-time teachers and mentors of Six Sigma. These people need to have good teaching, leadership, and quantitative skills. We suggest that most of the training for these people, beyond their Six Sigma Black Belt training, be through on-the-job training and experience. During the initial stages of a Six Sigma business strategy, external consultants can fill this role.

9.3   COMPUTER SOFTWARE

In today's world, statistical calculations are rarely performed by hand; rather, they are relegated to specialized statistical software packages. We believe that success with Six Sigma requires that practitioners have access to user-friendly and versatile statistical software. Such software should be made available to them at the beginning of their Six Sigma in-class training. The use of statistical software during training sessions expedites the learning of both the software and the statistical methods. It is also easier to master new software when others within the training are familiar with the software that is being described.

After training is complete, the availability and use of "company standard" statistical software enhances communications within and between organizations, including suppliers and customers, due to a common reference language. In our opinion, it is best if an organization is able to choose a company-wide standard statistical software package that offers an extensive array of tools, is easy to use, is competitively priced, and offers effective technical support. We cannot emphasize enough the need to ensure that a complete suite of Six Sigma statistical tools is available in one place. There are multiple statistical software packages available commercially or as freeware. However, most of them are not optimized for a Six Sigma approach to process improvement. Two journals for identifying commercially available, general-purpose statistical software are *Quality Progress* (Lundin, 1999) and *Quality Digest* (Floberg, 1999; Quality Sourcebook, 2000). The software package most frequently used by organizations that train Six Sigma practitioners is Minitab.

Reiterating, we believe that Six Sigma workshops should include only a minimal number of manual exercises, which are to convey basic understanding. Most of the instruction within an S^4 workshop should involve the use of computer software on a portable computer assigned to the individual. Ideally, students should own the software being used so that after the workshop sessions have concluded they will have the tools needed to efficiently apply S^4 techniques to their projects. Numerous studies have shown that information and skill acquired at training sessions are quickly lost if they are not immediately used on the job.

It should be emphasized that even though very powerful statistical software packages are readily available at reasonable cost, process improvements are not a direct result of efficient number crunching. Unfortunately, current computer programs do not give the Six Sigma practitioner the knowledge to "ask the right question" or apply the tools effectively. Access to appropriate statistical software is a necessary, but not sufficient, condition for meaningful bottom-line results.

9.4 INTERNAL VERSUS EXTERNAL TRAINING AND INVESTMENT

There is debate over the merits of in-house training versus training using external sources. Organizations often become confused when addressing this topic. Let us consider some of the issues.

Organizations interested in initiating an S^4 business strategy from scratch do not have off-the-shelf materials available to use even if they have qualified instructors on staff. Most people drastically underestimate the time and cost associated with developing training materials of the scope and complexity required for effective Six Sigma training. Developing a world-class course can take many years of effort. The achievement of an effective S^4 business strategy requires a much higher level of quality in instruction and materials than is provided by a typical business workshop. The advanced statistical material demanded by an S^4 approach to process improvement is complex and voluminous. It is essential that the materials be carefully designed and masterfully delivered. Most organizations, we believe, should not attempt to create their own Six Sigma training materials.

If an organization has a "corporate university," with qualified faculty and curriculum development specialists, it may be feasible to develop the material. However, for Six Sigma to be successful there is more involved than simply combining previous statistical process control (SPC), design of experiments (DOE), and other course material and calling it Six Sigma. Six Sigma courses should address not only the tools of Six Sigma but also organizational infrastructure support issues.

Another option is to hire a company to deliver Six Sigma training and coaching. This type of service can be very valuable during the initial stages of a Six Sigma rollout. Typically, the material can later be licensed for presentation by a core group of individuals within the company. Six Sigma training or coaching may initially appear expensive, but it is not nearly as expensive as trying to accomplish the task without having the resources and time to do so. A major portion of the initial investment in a Six Sigma business strategy is the time people devote to training and implementation. Consequently, it is not wise to cut corners by attempting to quickly develop Six Sigma in-house training approaches and material to fulfill training needs. When an initial Six Sigma training wave is not successful because of poor material or training, the organization's future chances of success with Six Sigma are decreased. This does not reflect well on those who are sponsoring the implementation.

10

PROJECT SELECTION, SIZING, AND OTHER TECHNIQUES

Using an S^4 methodology requires directing efforts toward projects having the most benefit to the customers, employees, and the bottom line of an organization. Defining the best problem to solve can often be more difficult than the analytical portion of a project. Many team leaders aren't sure their projects meet the critical needs of the organization. They find the process of choosing an appropriate project to be nebulous and frustrating. If the initial phase of a process-improvement project is not completed with care and rigor, project work can take many wrong directions. An effort must be made at the outset to *answer the right question.*

Organizations must develop a process to define strategic focus areas and then effectively scope them into manageable S^4 improvement projects. If a process is not identified and firmly established within an organization, and project selection is left to team leaders, the overall Six Sigma effort risks losing effectiveness. Projects may become isolated endeavors, and change may be small and distributed.

Successful projects are not chosen from thin air, or to meet an arbitrary deadline. The process begins with the executive team determining the strategic goals for the first wave of Six Sigma. Before team leaders can answer the right question, the leadership of the organization should identify operational objectives for each business unit and baseline the critical processes (Harry, 2000). To be truly effective, a strategic process for project selection, sizing, and execution should be established within an organization. The larger the company, the greater the need for this process. Figure 10.1 shows a general process map for selecting and executing effective Six Sigma projects.

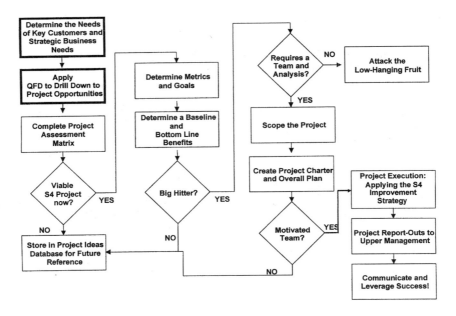

Figure 10.1 Process map of selecting effective Six Sigma projects

The goal of this chapter is to demonstrate effective methods to select, scope and execute successful Six Sigma projects. The sections are organized to follow the process map shown in Figure 10.1, each section highlighting important steps of an overall project-selection process. The first two steps (outlined in bold) were discussed in previous chapters.

10.1 PROJECT ASSESSMENT

Different companies have different strategic goals associated with Six Sigma. In a recent article, Jerome Blakeslee (1999) defines the essence of Six Sigma as "being able to examine and close the gap between what a business process produces and what customers demand." He then recommends that the width of the gap be used to prioritize Six Sigma efforts.

As mentioned in Chapter 8, "Creating a Successful Six Sigma Infrastructure," one of the methods executives or champions can use to begin to prioritize their Six Sigma efforts is the Customer House of Quality. To progress, Six Sigma practitioners from specific departments can take the output of the House of Quality process (i.e., a list of project opportunities) and brainstorm specific projects that address the strategic needs of the organization.

One successful method utilized at GE was to develop a list of critical-to-quality characteristics, or CTQs, to ensure that projects met customer needs and strategic business goals simultaneously. Every new project had to meet one of these strategic business goals. Examples of CTQ include delivery, cycle time, price value, and supplier quality. If a GE project didn't address one of those CTQs, it was archived for review at a future date. To further filter down to high-impact projects, the project savings had to exceed $50,000. Depending upon the strategic goals of a company, filters for the project selection process could include:

- Data availability
- Resource availability
- Estimated duration
- Meaningful defect definition
- Return on investment
- High probability of success
- Support of a strategic business goal or CTQ

Organizations might also consider having within their first wave of training a project filter of "easy to implement." Such a filter will make it more likely that the first project assigned for training is manageable and has a high likelihood of success. This will provide the Six Sigma practitioner with the experience and confidence needed to tackle more complex issues. It will also yield early success stories to share with employees, increasing the company's confidence in the S^4 methodology.

Guesswork can be taken out of the project selection process by using a project selection matrix as shown in Table 10.1, which is similar to the cause-and-effect matrix described in Breyfogle (1999), *Implementing Six Sigma* (chapter 13). This sample matrix relates to a company that defined "efficiency of their purchasing department" as a strategic goal for the first wave of their Six Sigma implementation. In this example, a Black Belt with sourcing experience was assigned to the group. During the first week of the assignment, the Black Belt and the manager, who was champion for the project, held a brainstorming session to determine the project she would take into training. They used a tool similar to the project-selection matrix and determined a project to focus on, having established that it would be best to work on a standardized process for purchasing first.

When creating a similar matrix within your organization, each column of this matrix should contain the critical criteria your company uses in selecting a Six Sigma project. The column headings should be predefined, with rankings appropriate for the needs of the business. It is then up to project sponsors and Black Belts to apply the project-selection matrix to opportunities in their functional areas and to develop a list of prioritized projects. By focusing on projects

TABLE 10.1   Project Assessment Matrix

Project Assessment Matrix	Complete within 6 months	Supports Strategic Business Goals	Meets Customer Needs	High Return on Investment	Motivated Team Available	Data Available	Meets Budget Constraints	Process Outputs
	10	10	8	7	7	6	6	◄ Importance
Projects								**Total**
Standardized Process for Purchasing	9	9	3	9	3	1	9	**348**
Improve Customer Service	9	3	9	9	3	1	9	**336**
Improve On-Time Delivery	9	1	9	3	1	9	9	**308**
Reduce Purchased Labor and Material Costs	3	9	3	9	3	1	9	**288**
Supplier Management	3	9	3	9	3	3	3	**264**
Improve Employee Training	1	3	3	9	9	1	3	**214**

that correlate to a list of project filters/criteria, resources will be most effectively applied to areas with the biggest impact on strategic goals. The mechanics of this simple cause-and-effect matrix are described in a later section of this chapter, called "Scope the Project."

To balance the effort of establishing strict project guidelines, consider creating a database to store project ideas that were not "big hitters" for the first wave of training. The collected ideas can be categorized appropriately and matched with strategic business goals for future waves. This also provides a method to capture project ideas from employees, increase buy-in to the Six Sigma business strategy, and assist teams brainstorming for new project ideas.

10.2   DETERMINE METRICS AND GOALS

Metrics drive behavior and require careful consideration prior to implementation. The selection of metrics and goals for quality project teams is crucial to the success of the teams and the company. Managers should assign performance goals that are demanding but flexible. Teams can then use creativity and become empowered into shaping their own metrics.

Project metrics should focus on measuring the process, not the product. Successful project metrics contribute to process knowledge, help to create a baseline for strategic processes, and are characterized by the following traits:

Customer Focused

The metrics used to evaluate the success of process improvement teams should relate to customer values, such as product quality, on-time delivery, and customer service. Reviewing the project selection House of Quality can be beneficial to a brainstorming session to create customer-centered metrics.

Cross-functional

Traditional measurements are functional; they were established to measure progress toward goals. When metrics fail using traditional measurements, a team is apt to forget its goals and return to the old "functional" way of working. Historically, most companies have not implemented cross-functional metrics except in the area of financial measures. A primary project metric should be cross-functional, helping teams clearly evaluate how well they are producing a desired result.

Informational

Metrics should be simple and straightforward, requiring a minimum of manipulation to become meaningful. Metrics should measure performance across time so that they can be analyzed for trends and improvement opportunities. Relative numbers are usually more informative.

Metrics should also create a common language among diverse team members. Metrics that are created by teams should have a purpose, buy-in, and empowerment. When creating project metrics, one should consider how these metrics link to key business metrics. Figure 10.2 shows how project selection metrics at the production and factory levels can relate to key business metrics. Typically, there is no one metric that fits all the requirements for a particular situation. Teams might conduct a brainstorming session to decide what metrics will help them to better their performance. The team can then review these metrics with executive management to ensure that they are in accord with the overall strategy of the business.

Care should be exercised in determining what is measured. Metrics should make sense to all team members and be easily captured and calculated. Metrics should not be set in stone; rather, they should be periodically revisited for value added to the process. Some organizations give focus only to the rate of defects at the end of a process, using a "one-size-fits-all" metric that may not make

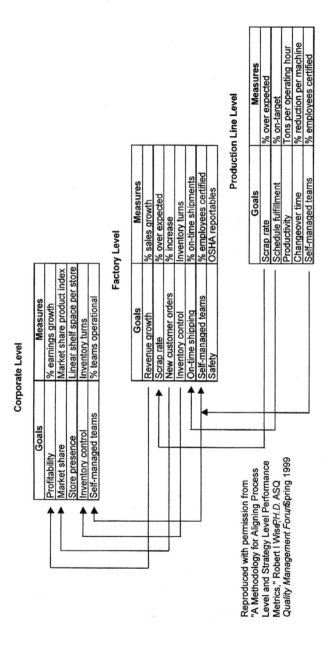

Figure 10.2 Selecting projects that link to key business metrics [Reproduced from Wise (1999), with permission]

sense. As discussed previously, this can happen when a sigma quality level metric is used to drive change. Measuring only defect rates or sigma quality level can lead to playing games with the numbers and ultimately turn a quality initiative into a bureaucratic game with numbers. In this situation, project leadership can degrade into an effort to make numbers meet artificial goals. With this approach, real change will be sporadic.

A balanced scorecard approach can be used within a Six Sigma business strategy. The balanced scorecard can address issues of quality, cycle times, and decreased costs. One balanced scorecard approach, described previously in this book, considers the categories of financial, customer, and internal business processes, and learning and growth. We believe that a balanced scorecard approach is needed; however, we suggest also considering reporting ROI values within summary reports whenever possible, even if assumptions need to be made when converting, for example, customer satisfaction to monetary units. The value of projects can be better understood when money is also used in reporting metrics. We believe that emphasis should be given to the search for smarter solutions that reduce the cost of poor quality (COPQ), as opposed to "playing games with the numbers," which often happens in Six Sigma programs.

Consider developing a project scorecard approach to metrics, since improvement in one area might degrade another. Figure 10.3 shows a sample project scorecard, including examples of measurements that an organization may want to reference when developing its own project metrics. Some items under "Customer" and "Learning and Growth" may seem difficult to measure. Consider using surveys for items that are nebulous and where objective data collection is difficult. The glossary in this book can be referenced for terms that are unfamiliar to the reader.

10.3   DETERMINE A BASELINE AND BOTTOM-LINE BENEFITS

In effective project selection, consideration should be given to how metrics are reported. Management should have systems in place to baseline significant business processes that are part of the Six Sigma strategic focus. Many tools exist that give a high-level view of a process. Consider using the following techniques to understand the capability of strategic processes:

Process Control Charts

As described previously, process control charts can be used to quantify process variability and identify when special cause occurs. A "30,000-foot level" control chart format can give a long-term point of view that reduces the amount of

Project Scorecard

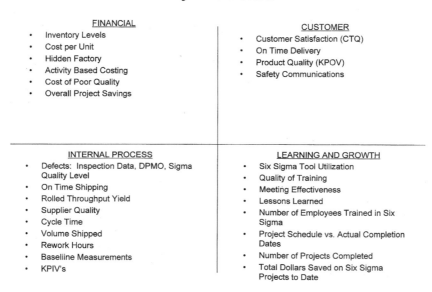

FINANCIAL	CUSTOMER
• Inventory Levels • Cost per Unit • Hidden Factory • Activity Based Costing • Cost of Poor Quality • Overall Project Savings	• Customer Satisfaction (CTQ) • On Time Delivery • Product Quality (KPOV) • Safety Communications
INTERNAL PROCESS	LEARNING AND GROWTH
• Defects: Inspection Data, DPMO, Sigma Quality Level • On Time Shipping • Rolled Throughput Yield • Supplier Quality • Cycle Time • Volume Shipped • Rework Hours • Baseliine Measurements • KPIV's	• Six Sigma Tool Utilization • Quality of Training • Meeting Effectiveness • Lessons Learned • Number of Employees Trained in Six Sigma • Project Schedule vs. Actual Completion Dates • Number of Projects Completed • Total Dollars Saved on Six Sigma Projects to Date

Figure 10.3   Project scorecard

firefighting within an organization and can help refocus common cause issues toward process improvement activities. Other statistical tools can then be used to identify the amount of common cause variability and what should be done differently to improve the process. For identified Key Process Input Variables (KPIVs), control charts can be created and utilized at a "50-foot level" to track the variables and identify when special causes occur so that immediate action can be taken.

Care should be taken in how data are collected. The effectiveness of control charts is highly dependent on the rational subgrouping that is used during data collection. Sampling plans can make a big difference in what is reported as special cause or common cause.

When creating "30,000-foot level" control charts, parts should be selected so that time between samples captures the entire variability of the process. In addition, a control chart strategy should be selected that uses subgroup variability when determining control limits. Breyfogle (1999), in *Implementing Six Sigma* (chapter 10), describes different control chart techniques and how an *XmR* chart, for example, can in many cases be a better charting alternative than a p chart or an $\bar{x}$ and R chart.

Histogram and Dot Plot

Histograms and dot plots give a picture of the data spread, but it is difficult to estimate the percentage that lies outside of the specification limits. For histograms, data must first be grouped in cells or classes.

Normal Probability Plot

An alternative to a histogram and dot plot, a normal probability plot can estimate the proportion of data beyond specification limits and can give a view of the "capability" of the process. This is an excellent tool when no specification limit exists for transactional/service processes. Normal probability plots also assess the validity of the normality assumptions.

Process Capability/Performance Indices

The purpose of process capability/performance indices is to quantify how well a process is executing relative to the needs of the customer. These indices give insight into whether defects are the result of a mean shift in the process or excessive variability. Example of process capability/performance indices are C_p, C_{pk}, P_p, and P_{pk}. Practitioners need to be careful in selecting the procedure they use to calculate and report these metrics. Breyfogle (1999), in *Implementing Six Sigma* (chapter 11), describes the confusion that exists over the definition of short-term vs. long-term variability and what equation/approach should be used to determine standard deviation for a given process. Care needs to be exercised when reporting these metrics, since everyone does not "speak the same language."

Once baseline metrics and charts are created, team leaders and upper management need to assess if something should be done differently. A calculation of the bottom-line benefit achievable through work in this area can be very beneficial. Calculations can include the cost of poor quality (COPQ), which can be used to determine the value of the project to the organization. The COPQ calculation could be considered the "cost of doing nothing."

It is a big leap to assume that quality improvements will correlate with profit improvements. Focusing on project savings alone can lead companies to overlook valuable projects that can have an indirect impact on customer satisfaction, with severe ramifications for future profit. These issues can also hurt the long-term prospects of Six Sigma methodologies within a company. Organizations need to consider a balanced approach to project selection when implementing Six Sigma.

10.4   SCOPE THE PROJECT

Many projects start out with goals and objectives that are too broad. Project leaders may feel overwhelmed with the task at hand. They may flounder, revisiting the measure phase and rethinking their project approach in frustration. This can wear on the confidence of the Six Sigma practitioner and performance levels of the team. This frustration often stems from the fact that leaders are working on a project that is too big and are actually trying to solve multiple projects at once. A strategy is needed to help organizations narrow the scope of projects and focus on areas that have the biggest impact first.

Frequently, the initial project goals that managers set for their teams serve as "parent projects," with multiple projects that roll up to meet the desired performance metrics. A method should be established to keep track of "child projects" to be completed at a future date so they are part of the overall plan for selecting projects.

If a project has been properly scoped, an organization should be able to answer the following questions when the project is finished:

- How much does this issue cost the business?
- What is the procedure to ensure quality products?
- How are results controlled?
- How much effort is given to this issue?
- How much improvement has been made in the last six months?
- What is the defect?
- What are our improvement goals?

Consider having project leaders fill out a checklist before submitting a project for approval. One section could deal specifically with project bounding, asking if the project has customer impact; does not try to "boil the ocean"; and can specifically answer "what is the defect?" An organization is on the right track if it can answer the questions listed above before establishing a project team and creating a charter.

Process Mapping

Creating a process flowchart as a team is a great way to display an accurate picture of the process and to gain insight into opportunities for improvement. Later in the project, the flowchart can be used for maintaining consistency of process application and subsequent improvement/establishment of standard operating procedures.

Process mapping can be laborious if the process is not bounded and the detail level agreed upon beforehand. The following are tips for proficient process mapping:

- Keep the initial view at a high level.
- Define subprocess steps that need further expansion off-line with other teams.
- Document the process as it currently exists.
- Document the dialogue that ensues during the creation of the flowchart.
- Document where the biggest bottlenecks occur.
- Note any steps that can be eliminated.
- Reference initial baseline measurements taken prior to process flowcharting to aid the team in determining at what detail to map the process.

The Six Sigma practitioner should embrace any conflict that arises during this process. The dialogue that emerges from a team during a flowcharting session is as important as the output of the session. Conflict in the discussion may pinpoint problems in the actual process. For further details and the mechanics of creating process flowcharts, refer to Chapter 4 of Breyfogle (1999), *Implementing Six Sigma*.

Pareto Chart

Pareto charts can help identify sources of common cause variation within a process. Pareto charts are based on the principle that a "vital few" process characteristics cause most quality problems, while the "trivial many" cause only a small portion of the quality problems. When nonconformance data are classified into categories, most of the time it is evident where initial project focus should lie.

Cause-and-Effect Diagram

When the process map is complete, it is time to narrow down the focus further. A brainstorming session can be held to determine the key causes that are creating defects within a process. A helpful tool in determining these causes is the cause-and-effect diagram.

When creating a cause-and-effect diagram, a team should keep in mind the following: The output of a cause-and-effect diagram depends to a large extent on the quality and creativity of the brainstorming session. A good facilitator can help teams stay creative and ensure that feedback is received from all participants.

One approach when conducting a brainstorming session is to have team members spend 10 to 15 minutes writing their ideas on notecards in silence. Next, shuffle the cards, reading each aloud as the cards are organized within similar categories in plain view of all team members. Blank notecards can then be used to capture the new ideas that will be created during the discussion of each card.

Once all new ideas have been submitted, spend some time reorganizing the cards and combining similar ideas. Often, facilitators use the following categories to group the data:

- Personnel
- Materials
- Measurements
- Methods
- Machines
- Environment

This approach sometimes helps teams to focus their thinking if the session is generating few ideas. However, it can stifle the creativity that goes into thinking about causes. We believe it is more effective to assign natural categories for the causes once brainstorming sessions are complete.

Cause-and-Effect Matrix

When the cause-and-effect diagram is complete, a method is needed to rank the list of causes and determine critical focus areas. A cause-and-effect matrix is a simplified QFD (quality function deployment) that gives focus to the inputs that are most likely to affect output (KPOVs).

In the cause-and-effect matrix shown in Table 10.1, outputs are listed in the top row and are assigned values according to importance to customers and strategic business goals. A higher number indicates more importance. The row contains a list of input variables from a process map or causes from a cause-and-effect diagram. At the intersection of each row and column, values are entered that quantify the amount of correlation thought to exist between the row and column factors. Relationships that are thought to be stronger can be given more weighting. One might, for example, use the following scale:

0 = no correlation
1 = remote correlation
3 = moderate correlation
9 = strong correlation

Numbers in each input row are cross-multiplied with the importance number at the top of the column and summed across each row. A ranking of the magnitude of the cross-multiplied values gives focus to the input variables the team thinks are most important. The highest-ranked values are then considered to be KPIVs, and should be given the most attention for additional study and/or process improvement efforts.

If importance ratings for a cause-and-effect matrix are not established, a four-block approach to rating an initial cause-and-effect diagram is an effective method for determining impact areas or to further scope the project. The procedure consists of two series of rankings: the first establishes how much *impact* the cause has on the effect (high or low). The second establishes the amount of work required to *implement* change (easy or hard). After causes are ranked, they are placed in one of the four blocks shown in the lower right corner of Figure 10.4. This example is a continuation of the purchasing project discussed above. In this case, the Six Sigma practitioner had no data to assess the relationship between KPIVs and KPOVs, so this rating system was used to further scope the project and identify "low-hanging fruit" that could be fixed immediately (causes with a rating of 1).

As demonstrated in the example above, the cause-and-effect diagram may produce some "low-hanging fruit" causes that can be addressed immediately. These changes need to be incorporated into the process and a new set of measurements collected and analyzed along with other data. Now that a process baseline has been established, the benefits from fixing this type of project will be visible and immediate. This method—identifying small successes throughout the duration of the larger project—is an excellent way to keep team motivation alive.

10.5   CREATE PROJECT CHARTER AND OVERALL PLAN

The project charter is an important activity of the planning process for any new team. It acts as the preliminary project plan and increases teamwork. Many teams skip over this important step because they view it as a "soft skill" and don't understand its value in helping to create a well-functioning team.

Perhaps the most beneficial outcome of a charter is team cohesion. Creating a charter as a team communicates goals and individual responsibilities clearly. It is a good team-building exercise and is an opportunity to get potential roadblocks out in the open for discussion. When developed as a group, a charter can create a sense of team ownership of the project.

The second major benefit of a project charter is that it clearly defines expectations between managers and team members, preventing delivery of mixed messages. A charter should be reviewed with upper management at a kickoff meeting to ensure that the overall project supports company strategy.

At a minimum, the project charter should document the following key items:

- Mission: what the team is expected to produce
- Basis: how the team mission connects with the strategic plan of the business

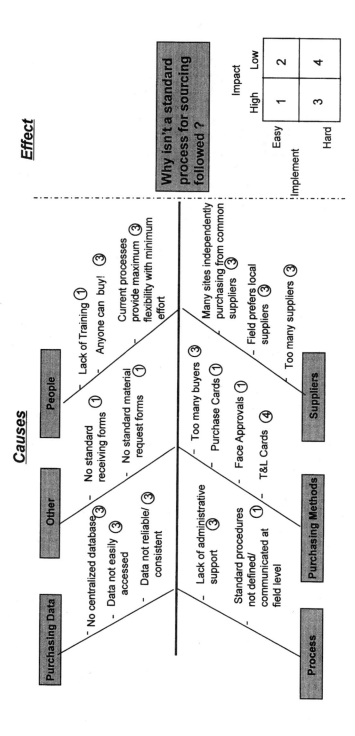

Figure 10.4 Cause-and-effect diagram

- Goals: performance challenges
- Roles: team members
- Responsibilities
- Project metrics: how the team will measure success
- Key milestones: report-out dates
- Key challenges: results of project risk-assessment activities
- Costs
- Benefits

A project charter should be updated as the project progresses so that it includes additional knowledge that has been obtained since the initiation of the charter. It is also a good idea to revisit the document when the team hits a roadblock or is lobbying for additional resources.

A sample project charter is shown in Table 10.2. It is presented with permission from Titanium Metals Corporation (TIMET), one of our clients, who utilized this charter to clearly define the roles and responsibilities of its newly established Six Sigma team. The team members are not like those of a conventional Six Sigma project; TIMET chose to supply additional resources to these highly visible projects in order to ensure success and create long-term buy-in to future Six Sigma activities.

TABLE 10.2   TIMET Project Charter

	PROJECT CHARTER
CHARGE	• Reduce rejections and rework attributed to surface cracks
START DATE	• November 15, 1999
FINISH DATE (for Improve Phase)	• June 30, 1999
GOAL	• Reduce cracks in Forge Products (specifics to be determined by inspection data and volume)
BENEFITS	• Improve FTP • Reduce Rework; Decrease Cycle Time • Reduce Past Dues; Improve On-Time Delivery • Total Savings: (to be determined upon completion of measure phase)

KEY MILESTONES

Define Phase Completion	• January 14, 2000
Measure Phase Completion	• March 10, 2000
Analyze Phase Completion	• May 5, 2000
Improve Phase Completion	• June 30, 2000
Control Phase Completion	• August 2000 (TBD)
Verify Phase Completion	• TBD

PROCEDURES
- TIMET Guidelines for Effective Groups
- "Red, Yellow, Green" Status Reports
- Six Sigma Methodology

SPONSOR
- Vice President of Information Technology and Strategic Change

ROLES AND RESPONSIBILITIES

Team Members

These Roles & Responsibilities also apply to all other team members with additional responsibilities identified below.

- Openly contribute process expertise and equipment knowledge
- Communicate changes with co-workers on project and solicit input
- Accept and complete action items on time
- Attend and participate in all meetings
- Collect data (as assigned)
- Treat other team members with respect and consideration
- Assist with DOE (as assigned)
- Implement improvements (as assigned)
- Eliminate rumors (open and honest communication)
- Communicate lessons learned from past process changes
- Identify "Cause and Effect"
- Follow Guidelines for Effective Groups
- Ensure process changes are sustained
- Seek out additional resources when needed
- Bring action items to the meetings
- Lead or participate in sub-team meetings as required

Team Leader
- Schedule & lead team meetings
- Direct team activities (Agendas, Assignments, Priorities)

- Issue weekly reports
- Present management report-outs
- Update Plant Manager (raise issues as necessary)
- Team stability & motivation
- Ensure Six Sigma is followed
- Interface with HR
- Interface with Union Officials

Co-Leader
- Identify risks & barriers
- Aid in data analysis
- Participate in developing priorities & tasks
- Provide metallurgical process knowledge
- Process observation & reporting
- Monitor Critical Success Factors

Black Belt
- Assist with management report-outs (statistics)
- Data analysis
- Oversees data collection
- Six Sigma Methodology Leader
- Select, teach, and use most effective tools
- Facilitate meetings (Adhere to Agenda)
- Identify Risks and Barriers
- Review Action-Item list at every meeting
- Distribute documents and update team members without e-mail

Project Manager
- Issue meeting minutes to all members
- Track and report major milestones
- Track and report detailed tasks
- Raise Red Flags in terms of schedule adherence
- Interface between Black Belts and finance

Process Owner–Production
- Communicate process knowledge
- Ensure that process improvements are implemented and sustained (Improve/Control phases)
- Communication & documentation of process changes
- Department team member participation (resolve scheduling conflicts)

Process Owner–Technical
- Communication & documentation of process changes
- Communicate metallurgical issues/knowledge & abate risks (with team)
- Obtain necessary approval for process changes (if required)
- Ensure that process improvements are implemented and sustained (Improve/Control phases)

Information Technology Analyst	• SAP data extraction, development & reports
	• Database management
	• Automation of metrics
	• Interface with BPO
Data Entry Coordinator	• Supports taking of minutes
	• Enters data into Excel and Access databases
	• Administrative Assistant
	• Meeting Logistics

Along with the project charter, a good plan is essential to the success of any project. Figure 10.5 shows a sample project plan, also from a team at Titanium Metals Corporation. They created this plan in the measure phase of their first Six Sigma project, and it served as a template for future project work. Again, this example is presented to show the items and detail included in an effective project plan. Duration dates for any particular project can differ significantly from these, depending on the scope of the project and team/management commitment, among other things. The ID nomenclatures within the figure as they relate to Six Sigma phases are as follows:

- Define phase: ID 1-8
- Measure phase: ID 10-23
- Analyze phase: ID 25-31
- Improve phase: ID 33-42
- Control phase: ID 44-49 (start and finish dates are TBD [to be determined])

When creating a plan, consider building the document in a team setting. Developing the plan as a team or reviewing it for feedback establishes buy-in and creates a vision of success for the future. One method we have had success with is to reference our 21-step integration of the tools when creating a detailed Gantt Chart for an S^4 Improvement project. This provides a structure in which to brainstorm.

10.6 MOTIVATED TEAM

To meet the challenges of today's business world, a team approach is needed. People want to work hard and make a difference, but individuals are limited in what they can achieve alone. It has been proven over and over again that team performance surpasses that of individuals acting alone.

During a postmortem, General Electric collected feedback showing that "human behavior was the single greatest cause of team-implementation failure." As discussed earlier, it is important to select full-time Black Belt candidates who not only have the ability to understand the statistical methodology but also possess excellent leadership and facilitation skills. To be truly successful, Six Sigma team members must be able to relate to each other as co-owners of the project. Relationship patterns in teams encompass the following six categories, ranging from adversarial to partnering (Weaver and Farrell, 1999):

1. **Coercion**:   "You must do this or face pain."
2. **Confrontation**: "You must do this."—"No, I won't!"
3. **Coexistence**:   "You stay on your side and I'll stay on mine."
4. **Cooperation**:   "I'll help you when my work is done."
5. **Collaboration**: "Let's work on this together."
6. **Co-Ownership**: "We both feel totally responsible."

A good facilitator can help a team progress into a less adversarial relationship. Weaver and Farrell describe a practical method for facilitators to use in promoting this growth (Weaver and Farrell, 1999). The approach, called TARGET (Truth, Accountable, Respect, Growth, Empowered, Trust), can be summarized as follows:

- Truth: Group members must know the truth about expectations and experiences. Facilitators can help create a climate of truth within group meetings by honestly reflecting their own experiences.
- Accountable: Group members need to be accountable to each other for their work. Facilitators can create a structure for accountability in meetings, such as reviewing action item lists and periodically revisiting the group's charter.
- Respect: Group members must experience others acting with integrity and honesty. Generating respect in a group is a tough task for a facilitator because it is so closely tied with trust. The best approach is to try to name the underlying problem directly and generate a plan to understand and address the issue.
- Growth: Team members and the team itself need to grow as a whole through learning to become more efficient at tackling difficult tasks. Facilitators can promote growth by helping groups recognize and value it.
- Empowered: Successful groups depend on independent action. Facilitators can help by identifying opportunities for independent action and can suggest methods for reflecting independent work back to the group.
- Trust: Effective groups have team members who trust each other and can count on each other to complete the necessary work of the team. Facilitators can help teams identify areas where trust is the strongest and the weakest.

ID	Task Name	Duration	Start	Finish
1	Clarify business goals and objectives of project	3d	Mon 11/15/99	Wed 11/17/99
2	Define objective & scope of project	4d	Wed 11/17/99	Mon 11/22/99
3	Identify team members	5d	Mon 11/29/99	Fri 12/3/99
4	Plan overall project	5d	Mon 12/6/99	Fri 12/10/99
5	Create project charter	6d	Mon 12/13/99	Mon 12/20/99
6	Obtain union support for project	1d	Mon 12/20/99	Mon 12/20/99
7	Establish measure phase project plan milestones	3d	Mon 12/20/99	Wed 12/22/99
8	Present project proposal to executive staff for approval	1d	Thu 12/23/99	Thu 12/23/99
10	Review & confirm standard operating procedures SOP	15d	Mon 1/3/00	Fri 1/21/00
11	Complete detailed process maps	15d	Mon 1/3/00	Fri 1/21/00
12	C&E Diagram with KPIVs	10d	Mon 1/10/00	Fri 1/21/00
13	Conduct benchmarking exercise	9d	Mon 1/10/00	Thu 1/20/00
14	Data collection activities	36d	Fri 1/21/00	Fri 3/10/00
15	Evaluate measurement system	41d	Fri 1/21/00	Fri 3/17/00
16	FMEA risk analysis with abatement plans	25d	Mon 2/14/00	Fri 3/17/00
17	Develop project scorecard with key project metrics	10d	Mon 3/6/00	Fri 3/17/00
18	Create process control charts	3d	Mon 3/13/00	Wed 3/15/00
19	Determine process capability	3d	Wed 3/15/00	Fri 3/17/00
20	Confirm project scope, goals, & financial benefits	2d	Fri 3/17/00	Mon 3/20/00
21	Six Sigma training for core team members	5d	Mon 3/20/00	Fri 3/24/00
22	Prepare analysis phase milestones with due dates	1d	Mon 3/27/00	Mon 3/27/00
23	Prepare presentation and report out	1d	Fri 3/31/00	Fri 3/31/00

25	Revise cause and effect analysis	3d	Mon 4/3/00	Wed 4/5/00
26	Analyze data for sources of variation	20d	Mon 4/3/00	Fri 4/28/00
27	Identify significant causal factors	11d	Fri 4/14/00	Fri 4/28/00
28	Determine additional data needed	1d	Fri 4/28/00	Fri 4/28/00
29	Six Sigma training for core team members	5d	Mon 4/24/00	Fri 4/28/00
30	Prepare improve phase milestones with due dates	1d	Wed 5/3/00	Wed 5/3/00
31	Prepare presentation and report out	5d	Mon 5/1/00	Fri 5/5/00
33	Develop critical path plan	1d	Mon 5/8/00	Mon 5/8/00
34	Complete DOE	24d	Tue 5/9/00	Fri 6/9/00
35	Six Sigma training for core team members	5d	Mon 5/22/00	Fri 5/26/00
36	Implement process changes	5d	Mon 6/12/00	Fri 6/16/00
37	Conduct risk assessment	5d	Mon 6/12/00	Fri 6/16/00
38	Design control tools like SPC	5d	Mon 6/19/00	Fri 6/23/00
39	Collect & analyze data to validate improvement	5d	Mon 6/19/00	Fri 6/23/00
40	Develop plan to quanitfy & fix causes of variation	3d	Wed 6/21/00	Fri 6/23/00
41	Prepare control phase milestones with due dates	3d	Wed 6/21/00	Wed 6/23/00
42	Prepare presentation and report out	5d	Mon 6/26/00	Mon 6/30/00
44	Conform improvements	1d	Mon 7/3/00	Mon 7/3/00
45	Put controls in place	1d	Mon 7/3/00	Mon 7/3/00
46	Monitor and control variation	1d	Mon 7/3/00	Mon 7/3/00
47	Establish transition plan	1d	Mon 7/3/00	Mon 7/3/00
48	Complete project	1d	Mon 7/3/00	Mon 7/3/00
49	Prepare presentation and report out	1d	Mon 7/3/00	Mon 7/3/00

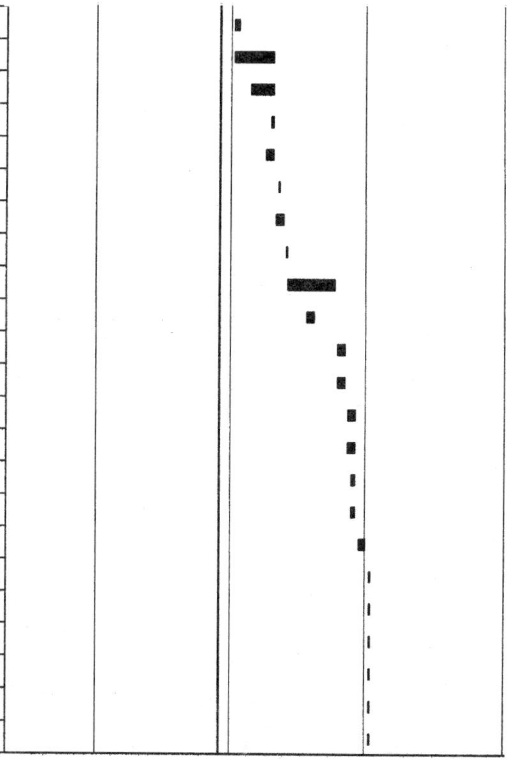

Figure 10.5 Gantt chart

General Electric learned that high-performance teams possessed the following elements: management support, shared vision, positive team members, and key stakeholders as team members. Other common traits of high-performance teams include the following:

Cross-Functional

Teams should have cross-functional representation and possess complementary skills. It is best to have only a few participants within each team. When a cross-functional team is assembled, most members have hidden agendas, depending on the department they are from. As a result, a professional facilitator is sometimes needed to keep a team on track and focused on one agenda. Facilitators can also help improve communications between departments and elicit feedback from reluctant team members.

Performance Challenges

A team's likelihood of success increases if its leaders clearly communicate challenging performance objectives based on the needs of customers, employees, and shareholders.

Meaningful Goals

Arbitrary goals set for individuals or organizations who have no control over the outcome can be very counterproductive and very costly. Each team member must feel that the purpose of the team is worthwhile and important to the organization as a whole. Successful teams have a shared purpose that can be articulated by all members.

Clear Approach

Team members must speak a common language. They need to understand and agree upon the approach to solving the problem at hand. Every member must understand the process they are undertaking.

Open Communications

Open, honest, and frequent communication among team members should be a primary goal. Members should feel free to speak their mind and give honest feedback.

Feedback

What gets measured and rewarded gets done, but few companies recognize or reward good performers effectively. To be truly motivating, the recognition process needs to be well thought out and personal. It should be done in a way that encourages cooperation and team spirit, such as at a company party or another forum that allows for public expressions of appreciation. Effective feedback can often be as simple as saying thank you.

Phillip Crosby Associates recognizes its employees with a "PCA Beacon of Quality" award (Crosby, 1995). The award is given to three individuals, chosen by their peers, who exemplify quality in the way they approach their work. The company finds it to be an excellent method to motivate employees—much more effective than, say, distributing token gifts or money awards.

10.7   PROJECT EXECUTION CONSIDERATIONS

Pressure on individuals and teams to produce can be enormous. A systematic approach to solving problems is often viewed as a waste of precious time. These feelings can lead to rushing, a common point of failure in many Six Sigma projects. As a result, teams are forced to reconsider a number of problems because their solutions were inadequate. Time spent at the outset defining the problem, collecting good data, ensuring that the data are telling a good story and listening to the story, can avoid this outcome and save the company from other problems down the road, not only in productivity but also in credibility.

When Six Sigma projects were first executed at General Electric, a measure-analyze-improve-control methodology was followed. The projects met with a good deal of success, but there were problems too. Project problems were most often related to not having:

- empowered, dedicated teams
- a project charter
- motivated team members
- scoped projects that related to key business strategies

It was noted that many projects tried to take on a "solving world hunger" mind-set and needed to be reworked into manageable sizes. Also, other projects seemed to be picked out of the blue and didn't have substantial benefit potential. Because of this problem, a phase to define projects was added, which resulted in the define-measure-analyze-improve-control (DMAIC) nomenclature for the overall Six Sigma implementation process. There was also a refocus on the training of Black Belts and Champions in project scoping, which significantly improved the success rate of projects.

When a project is properly scoped, sometimes only very simple changes are needed to make the process much better. Strict process improvement tactics that are applied rigidly can yield benefits from the "low-hanging fruit" that is prevalent on many projects at the start; however, "big hitters" in the process, those that are further up the tree, can require much creative thinking. Many initial projects can be simple reductions of unplanned slack and redundancy. But when tackling more persistent process problems, the statistical tools of a Six Sigma methodology can provide a wide array of options.

There are many combinations of Six Sigma tools and charts that will help determine the root cause of variation within processes. However, care must be taken to avoid becoming overwhelmed by the data and slipping into a "data analysis paralysis" mode. Also, many companies need to "clean up" their data collection and measuring processes before data can be analyzed in a meaningful way. For example, in many cases a Gage R&R study needs to be conducted early when investigating process improvement opportunities. Many companies initially find that their measurement systems are not accurate and require modification.

Figure 10.6 illustrates the concept of passive data collection of input variables that were thought to be significant from wisdom-of-the-organization assessments during the measurement phase of an S^4 project, using such tools as a cause-and-effect diagram, a cause-and-effect matrix, and FMEA. Subsequent activity, part of the analysis phase, can involve graphical tools such as multi-vari charts, box plots, and marginal plots. Testing significance and making predictions involve statistical tools such as regression, analysis of variance, correlation studies, and variance components.

Consider the possibility that the measure and analyze phases did not yield significant improvements through passive assessment. This could occur because the input variables were not large enough to detect a difference or because care was not taken to establish a data collection system that produced meaningful results. A DOE factorial methodology can be used to assess the impact of changes to the process. This approach involves making proactive changes for the purpose of determining what long-term changes are most beneficial for process improvement.

Six Sigma training provides a toolbox of statistical techniques and excellent problem-solving/project management skills that when combined can truly make breakthrough improvements in business processes. This approach can facilitate creativity and the willingness of individuals to let the "good" data lead them down the path to appropriate improvements. Again, the point of failure in many Six Sigma projects is rushing to solutions. Repeatedly, we see practitioners begin a project with the answer in mind. They go through the motions of applying the Six Sigma tools, but often they do not address the biggest opportunities for improvement and bottom-line benefits.

In an earlier chapter we discussed Six Sigma deployment. Figure 7.2 highlights our suggested overall business strategy of this deployment through projects. In the chapters that follow, we give specific examples that follow our "21-Step Integration of the Tools," a road map of how individual tools fit into the big picture.

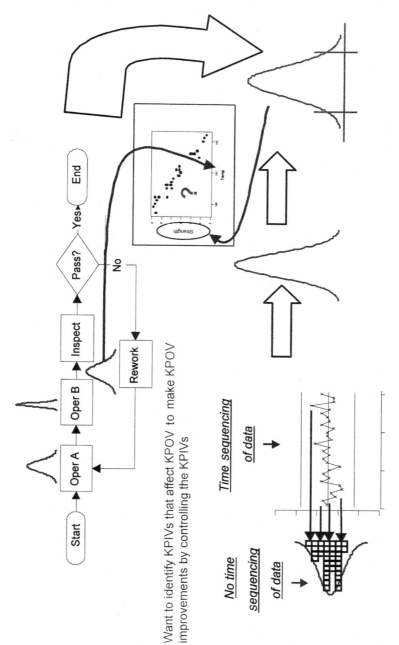

Figure 10.6 The determination of KPIVs that can affect KPOVs

173

10.8   PROJECT REPORT-OUTS TO UPPER MANAGEMENT

Six Sigma practitioners often feel pressured to apply all the tools they have learned and include them in presentations to executive management. This can result in a presentation that is difficult to follow.

Finding the root cause of a problem requires diligence and forethought. A cursory view of data is not sufficient, because it can lead to detecting mere symptoms of the true underlying problem. To be efficient, data analysis must use sophisticated statistical tools for the purpose of determining a concise and simple definition of the problem's key causes. It should tell a story that is well thought out and ends with a simple conclusion. An effective presentation can consist of as little as a few good charts.

We suggest that companies investigate how they can establish a periodic review process for all projects. When creating this "project report-out" process, a company should consider holding a brainstorming session with employees who will participate in all areas of the process. The goal is to obtain group consensus on how frequent project report-outs should take place and who should be involved.

We also suggest that the team spend time developing templates that Black Belts can follow in presenting their material to upper management. A report-out can take the form shown in Figure 10.7 through Figure 10.11, which describe how someone might present the work of a project to management in a closeout report. Graphs that tell the best story were selected from each phase.

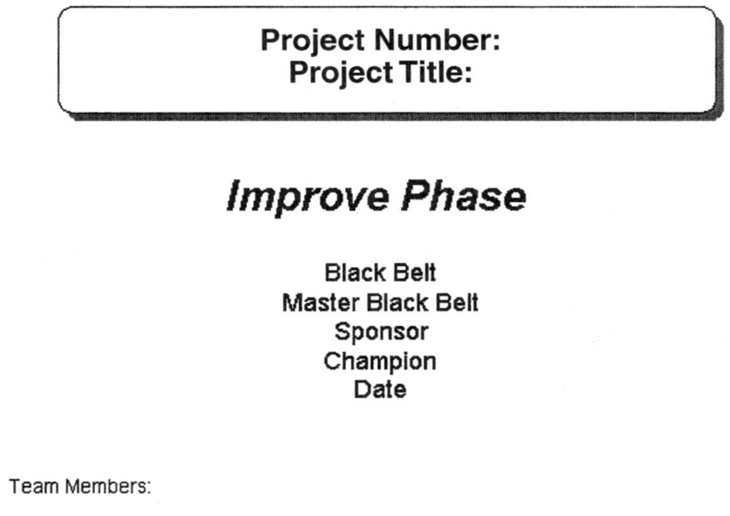

Project Number:
Project Title:

Improve Phase

Black Belt
Master Black Belt
Sponsor
Champion
Date

Team Members:

Figure 10.7   Project report-out: slide 1

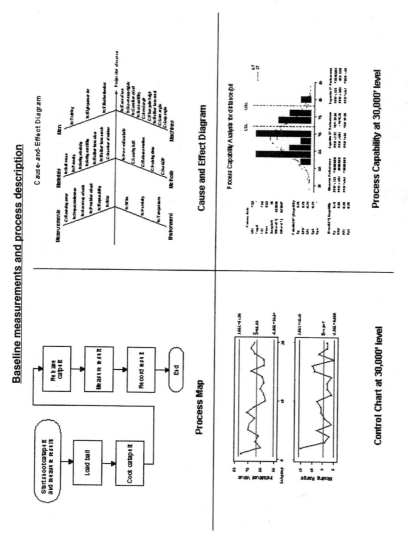

Figure 10.8 Project report-out: slide 2

175

Factors that were not going to be considered were removed
Importance measures were 0, 1, 3, 5, 9

Process Outputs >>>>>>>	Projection Distance (Average)	Projection Distance (Variability)	Cycle Time	Lateral Distance From edge of tape measurer	<<<<Process Outputs
Importance >>>>>>>>>>>>	10	10	5	2	<<<<<<<<Importance
---- Process input ----	------- Correlation of Input to Output -------				------- Total -------
Stop Angle	9	1	0	0	100
Start Angle	9	1	0	0	100
Hold Time	1	3	5	0	65
Release Method	1	5	0	0	60
Loading Ball	1	3	0	0	40
Measurement Rounding Error	3	3	0	0	60
Measurement Impact Reference Point	3	3	0	0	60

Used RPN of 200 as cut-off for action

POTENTIAL
FAILURE MODE AND EFFECTS ANALYSIS

FMEA Type (Design or Process):					Project Name/Description:							Date (Orig.):			
Responsibility:						Prepared By:						Date (Rev.):			
Core Team:												Date (Key):			
Design FMEA (Item /Function) Process FMEA (Function/Requirements)	Potential Failure Mode	Potential Effect(s) of Failure	S e v	C l a s s	Potential Cause(s) / Mechanism(s) of Failure	O c c u r	Current Controls	D e t e c	R P N	Recommended Actions	Responsibility & Target Completion Date	Actions Taken	S e v	O c c u r	D e t e c	R P N
Start Angle	Wrong setting	mean distance off target	7		Operator set wrong	6	None		9	378	SOP/DOE					
			7		Locking mechanism slips	5	None		9	315	SOP					
	Unstable	distance variability excessive	3		Wobble of holding mechanism	9	None		9	243	Redesign					
			3		Compressability of tubing	5	None		9	135						
		failure of holding mechanism	10	*	Locking mechanism screw not tight	3	None		9	270	SOP/Redesign					

Figure 10.9   Project report-out: slide 3

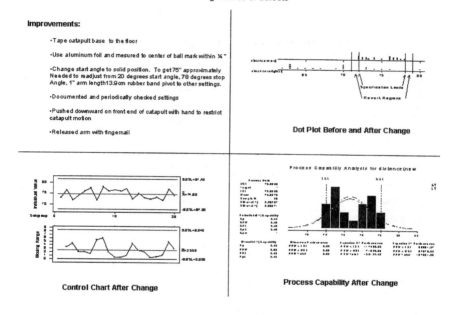

Figure 10.10   Project report-out: slide 4

Estimating Annual Financial Benefits

	Baseline	After Change
Inspection costs =	10	10
Rework cost =	100	100
Scrap cost =	200	200
Annual volume =	100,000	100,000
Median of data =	70	74
Upper spec limit =	73	77
Lower spec limit =	67	71
Rework upper limit =	74	78
Rework lower limit =	66	70
Number inspected =	100	100
Number to rework =	7	9
Number to scrap =	42	5
% reworked =	7%	9%
Annual cost of rework =	700000	900000
% scrapped =	42%	5%
Annual cost of scrap =	8400000	1000000
COPQ *	$10,100,000	$2,900,000
	$7,200,000	
* Includes inspection costs		

Figure 10.11   Project report-out: slide 5

In this example we use data collected from catapult exercises used in our workshops to illustrate how the Six Sigma tools can be integrated to solve project problems using an S^4 road map, as described in Figure 7.2. This report-out focuses on the "low-hanging fruit" fixes that the team completed initially; therefore, the report does not represent the full tool usage that project report-outs might contain.

The details of the catapult exercises are described in Breyfogle (1999), *Implementing Six Sigma*. The flow of tool usage and the people involved at various points is similar to what is shown in section 11.1.

Other slides to consider in a presentation of basic topics are the following:

- Solution Tree (e.g., Figure 10.12)
- Project Schedule
- Project Charter
- Results of Measurement System Analysis

10.9 COMMUNICATING AND LEVERAGING SUCCESS

As a mentor of Black Belts at General Electric, one of our authors (Becki Meadows) found that many of the GE project leaders were working on similar projects and often had the same frustrations. There was a lack of synergy between Black Belts and no established method for them to communicate. The need for mentoring and communicating lessons learned led to the creation of a quality web page. A new intranet site was created, containing the following elements:

- Templates for project execution
- Presentations of completed projects by theme or KPOV
- Explanations of the effective use of Six Sigma tools
- Lessons learned from experienced Black Belts
- Contests for "project of the month"
- A project selection process that maps steps, including for leveraging
- Templates for project report-outs

Coordination of quality projects through intranet sites or similar means is a necessary function in most organizations, one that is best accomplished by a dedicated executive leader or quality leader. If this function is overlooked, a company can waste resources in completing similar projects and constantly "reinventing the wheel." In subsequent training and project selection waves, it can be beneficial to include templates and leverage projects.

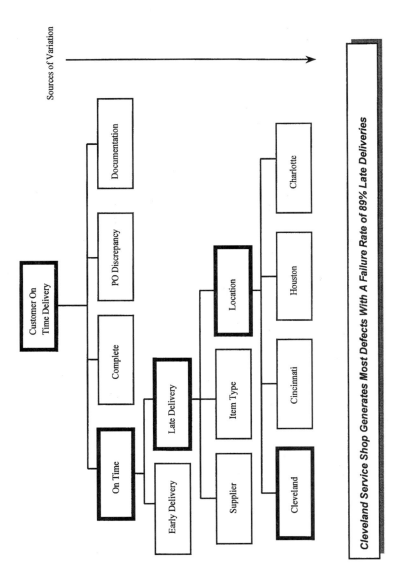

Figure 10.12 Solution tree example

PART 4

APPLYING SIX SIGMA

11

MANUFACTURING APPLICATIONS

Six Sigma techniques are applicable to a wide variety of manufacturing processes. The following examples illustrate how Smarter Solutions has assisted customers in benefiting from the wise application Six Sigma tools.

- *Customer 1 (Computer Chip Company):* Developed a control charting measurement system and a DOE strategy to improve the consistency of peel-back force necessary to remove packaging for electrical components. Results: The approach directly assesses customer needs by considering both within and between product variability.

- *Customer 2 (Computer Chip Company):* Developed a technique to monitor and reduce the amount of damage from control handlers. Results: The control chart test strategy considered sampled continuous data information to monitor the existing process in conjunction with DOE techniques that assess process improvement opportunities.

- *Customer 3 (Plastic Molding Company):* Applied DOE techniques to an injected molded part. Results: Part-to-part variability decreased. Company now routinely uses DOE techniques to better understand and improve processes.

- *Customer 4 (Computer Company):* Used DOE techniques to reduce printed circuit board defects. Results: Determined the manufacturing process settings that reduced the number of bridges, insufficient solders, and nonwet conditions.

- *Customer 5 (Environmental Equipment Company):* Used a DOE strategy to improve manufacturing product yield. Results: A small number of experimental trials uniquely assessed design and manufacturing parameters in conjunction with other factors that could affect test accuracy and precision.
- *Customer 6 (Telephone Equipment Company):* Implemented DOE techniques to improve the manufacturing of a critical plastic component. Results: The analysis approach identified important factor settings that improved the process capability/performance for flatness.
- *Customer 7 (Satellite Equipment Interface Company):* Developed a field reporting methodology. Result: The field tracking system quantifies product failure rate as a function of time and identifies the best area to focus improvement efforts.
- *Customer 8 (Video Conferencing Company):* Created a strategy to determine where to focus improvement efforts for product burn-in. Result: A multiple control chart strategy of both the burn-in and field failure rates is combined with Pareto charting to identify common and special causes along with opportunities for improvement.

These examples illustrate the many Six Sigma implementation strategies and unique tool combinations available for manufacturing processes. The techniques enable organizations to achieve significant benefits when they can extend their measurement and improvement activities from single products to a manufacturing process that creates various products. In addition, Six Sigma provides a means of exposing the hidden factory of reworks and helps quantify the cost of doing nothing. Through the wise application of Six Sigma techniques, organizations can focus on the big picture and then drill down when appropriate. This chapter includes thoughts on what can be done when measuring manufacturing processes, an overall application strategy, and application examples.

11.1   21-STEP INTEGRATION OF THE TOOLS: MANUFACTURING PROCESSES

There are many possible alternatives and sequencing patterns of Six Sigma tools for S[4] manufacturing projects. Our chart called "21-Step Integration of the Tools: Manufacturing Processes," which follows, is a road map to how individual tools fit into the larger manufacturing picture. Specific tools are highlighted in bold print to aid the reader in later locating where a tool may be applied. The glossary at the back of this book will help clarify terms. Breyfogle (1999), *Implementing Six Sigma*, provides details on the mechanics of using the tools.

21-Step Integration of the Tools: Manufacturing Processes

Step	Action	Participants	Source of Information
1	Identify critical customer requirements from a high-level project measurement point of view. Define the scope of projects. Identify **KPOVs** that will be used for project metrics. Establish a **balanced scorecard** for the process that considers also **COPQ** and **RTY** metrics.	Six Sigma practitioner and manager championing project	Organization wisdom
2	Identify team of key "stakeholders" for project. Address any project format and frequency of status reporting issues.	Six Sigma practitioner and manager championing project	Organization wisdom
3	Describe business impact. Address financial measurement issues of project.	Six Sigma practitioner and Finance	Organization wisdom
4	Plan overall project. Consider using this "21-Step Integration of the Tools" to help with the creation of a project management **Gantt chart**.	Team and manager championing project	Organization wisdom
5	Start compiling project metrics in time series format with sampling frequency reflecting "long-term" variability. Create **run charts** and **"30,000-foot level" control charts** of KPOVs. Control charts at this level can reduce amount of "firefighting."	Six Sigma practitioner and team	Current and collected data
6	Determine "long-term" **process capability/ performance** of KPOVs from "30,000-foot level" control charts. Quantify nonconformance proportion. Compile nonconformance issues in **Pareto chart** format.	Six Sigma practitioner and team	Current and collected data
7	Create a **process flowchart/ process map** of the current process at a level of detail that can give insight to what should be done differently.	Six Sigma practitioner and team	Organization wisdom

8 Create a **cause-and-effect diagram** to identify variables that can affect the process output. Use the **process flowchart** and **Pareto chart** of nonconformance issues to help with the identification of entries within the diagram.	Six Sigma practitioner and team	Organization wisdom
9 Create a **cause-and-effect matrix** assessing strength of relationships between input variables and KPOVs. Input variables for this matrix could have been identified initially through a **cause-and-effect diagram**.	Six Sigma practitioner and team	Organization wisdom
10 Conduct a **measurement systems analysis**. Consider a **variance components analysis** that considers repeatability and reproducibility along with other sources such as machine-to-machine within the same assessment.	Six Sigma practitioner and team	Active experimentation
11 Rank importance of input variables from the **cause-and-effect matrix** using a **Pareto chart**. From this ranking, create a list of variables that are thought to be KPIVs.	Six Sigma practitioner and team	Organization wisdom
12 Prepare a focused **FMEA**. Consider creating the FMEA from a systems point of view, where failure mode items are the largest-ranked values from a **cause-and-effect matrix**. Assess current **control plans**.	Six Sigma practitioner and team	Organization wisdom
13 Collect data for assessing the KPIV/KPOV relationships that are thought to exist.	Six Sigma practitioner and team	Collected data
14 Use **multi-vari charts, box plots**, and other graphical tools to get a visual representation of the source of variability and differences within the process.	Six Sigma practitioner and team	Passive data analysis
15 Assess statistical significance of relationships using **hypothesis tests**.	Six Sigma practitioner and team	Passive data analysis

16 Consider using **variance components analysis** to gain insight into the source of output variability. Example sources of variability are day-to-day, department-to-department, part-to-part, and within part.	Six Sigma practitioner and team	Passive data analysis
17 Conduct **correlation, regression**, and **analysis of variance** studies to gain insight into how KPIVs can impact KPOVs.	Six Sigma practitioner and team	Passive data analysis
18 Conduct factorial **DOE**s and **response surface analyses**. Consider structuring the experiments so that the levels of KPIVs are assessed relative to the reduction of variability in KPOVs. Consider structuring the experiment for the purpose of determining KPIV settings that will make the process more robust to noise variables such as raw material variability.	Six Sigma practitioner and team	Active experimentation
19 Determine optimum operating windows of KPIVs from **DOE**s and other tools.	Six Sigma practitioner and team	Passive data analysis and active experimentation
20 Update **control plan**. Implement "50-foot level" **control charts** to timely identify special cause excursions of KPIVs.	Six Sigma practitioner and team	Passive data analysis and active experimentation
21 Verify process improvements, stability, and **capability/ performance** using demonstration runs. Create a final project report stating the benefits of the project, including bottom-line benefits. Make the project report available to others within the organization. Monitor results at 3 and 6 months after project completion to ensure that project improvements/benefits are maintained.	Six Sigma practitioner and team	Active experimentation and/or collected data

The following examples highlight how to utilize the steps outlined above to gain insight into and improvement in manufacturing processes.

11.2   EXAMPLE 11.1: PROCESS IMPROVEMENT AND EXPOSING THE HIDDEN FACTORY

A company assembles surface mount components on printed circuit boards. This company might report their defective failure rate in the format shown in Figure 11.1. This type of chart is better than a run chart where there is a criterion, which could induce the firefighting of common cause variability as though it were special cause. The centerline of this control chart can be used to estimate the capability of this process, which is 0.244. Given that a 24.4% defective rate is considered unacceptably high, a Pareto chart like the one shown in Figure 11.2 can give insight as to where process improvements should be focused. From this chart it becomes obvious that the insufficient solder characteristic should be attacked first.

A brainstorming session with experts in the field can then be conducted to create a cause-and-effect diagram for the purpose of identifying the most likely sources of the defects. Regression analysis followed by a DOE might then be appropriate to determine which of the factors has the most impact on the defect rate. This group of technical individuals can perhaps also determine a continuous response to use in addition to or in lieu of the preceding attribute response consideration. One can expect that a continuous response output would require a much smaller sample size.

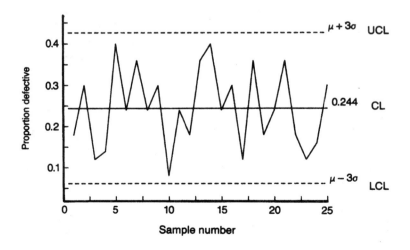

Figure 11.1   p-chart of defective assemblies

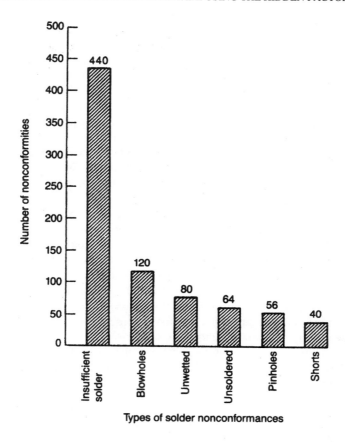

Figure 11.2   Pareto diagram of solder defects [From Messina (1987), with permission]

After changes are made to the process and improvements are demonstrated on the control charts, a new Pareto chart can be created. Perhaps the improvements to the insufficient-solder noncompliance characteristic will be large enough that blowholes will be the largest of the "vital few" to be attacked next. Process control charts can also be used to track insufficient solder in isolation so that process degradation from the "fix level" can be identified quickly.

Improvements should be made to the manufacturing process after a confirmation experiment verifies the changes suggested by the experiments. Because of these changes, the data pattern of the control chart should now shift downward in time to another region of stability. As part of a continuing process improvement, the preceding steps can be repeated to identify other areas for improvement.

This approach is one way to define the process measurement and improvement strategy. However, often in this type of situation each of the company's manufacturing lines assembles various product types, which consist of differing circuit layouts, component types, and number of components. For this situation one might wish to create a measurement system and improvement strategy as described above for each product type. A major disadvantage to the approach described above is that emphasis is given to measuring the product versus measuring the process. Since most manufacturing improvements are made through adjustments to the process, as opposed to the product, a measurement system is needed that focuses on the process and what can be done to improve it. A couple of Six Sigma metrics may prove to be very valuable for this situation; these metrics are DPMO and rolled throughput yield.

Let us consider first DPMO. This metric can also bridge product type and expose the hidden factory of reworks. As described earlier, the defects per opportunity (DPO) calculation can give additional insight into a process because it accounts for number of opportunities for failure. Consider, for example, that there were 1,000 opportunities for failure on an assembly, consisting of the number of solder joints and the number of assembly components. If 23 defects were found within the assembly process after assembling 100 boards over a period of time, the estimated ppm rate for the process for that period of time would be 230 (i.e., $1,000,000\{23/[100 \times 1,000]\} = 230$). If someone were to convert this ppm rate to a sigma quality level, they would report a 5.0 sigma quality level for the process using Table 2.1. The sigma quality level of 5 initially seems good, since average companies are considered to be at a 4.0 sigma quality level, as noted in Figure 2.3. However, if these defects were evenly distributed throughout the 100 assemblies produced such that no more than one defect occurred on each assembly, 23% ([23/100]100) of the assemblies would exhibit a defect somewhere within the manufacturing process. We believe most readers consider this an unsatisfactory failure rate for the manufacturing process, which is one reason why we discourage the reporting and use of the sigma quality level metric to drive improvement activities.

However, a reported ppm rate can be a very beneficial metric in that it can create a bridge from product to process and can expose the hidden factory of reworks. We should note that it would be difficult to examine the magnitude of this metric alone to quantify the importance of project activities within this area. For example, the manufacturing of a computer chip might have millions of junctions that can be considered opportunities for failure. For this situation, we might report a sigma quality level much better than *six* and still have an overall defective rate of the assembled unit of 50%.

Another assembly situation might have only a few parts that are considered opportunities for failure. For this situation we might report a sigma quality level of 4.0 but an overall defective rate of the assembled unit of 0.1%. Because of this discrepancy we suggest tracking ppm rates directly over time using a control chart. When a process exhibits common cause variability, the centerline of the control chart can be used to describe the capability/performance of the process. Cost-of-poor-quality calculations can then be made using this value and the cost associated with the types of defects experienced in the manufacturing process. This cost of poor quality can be used to determine if a project should be undertaken, or as a baseline for determining the value of process improvement activities resulting from a project.

Rolled throughput yield is another metric that can be useful in this situation. This metric, which quantifies the likelihood of product progressing through a process without failure, can expose major sources of reworks within a factory. To create this metric, the process is subdivided into process steps such as pick-and-place of components, wave solder, and so forth. The yield for each step is multiplied together to determine rolled throughput yield. Steps that have low yields are candidates for initial S^4 process improvement efforts.

Other areas of this manufacturing process that could be considered for projects are inventory accuracy, frequency and cause of customer failures, expense reduction, effectiveness of functional tests, and functional test yields.

11.3   EXAMPLE 11.2: BETWEEN- AND WITHIN-PART VARIABILITY

Chapters 10, 34, and 35 of Breyfogle (1999), *Implementing Six Sigma*, describe a variety of control charting techniques. The example described here is an application of the three-way control chart and improvement efforts.

Plastic film is coated onto paper. Three samples are taken at the end of each roll (Wheeler, 1995a) in such a way that the time between samples captures the entire variability in the process. Care needs to be exercised when evaluating this type of situation. Sometimes practitioners erroneously create $\bar{x}$ and R control charts using within-part variability as though they were measurements from different part samples. This type of data should be plotted as a three-way control chart, which can then be used to understand the variability between and within rolls, as shown in Figure 11.3.

The chart for subgroup means from this figure shows roll-to-roll coating weights to be out of control. Something is apparently happening within this process to allow film thickness to vary excessively. The sawtooth pattern within the moving range chart suggests that there is a large change in film thickness every other roll. The sample range chart is indicating stability between rolls; however, the magnitude of this positional variation is larger than the average moving range between rolls, which is an opportunity for improvement.

We are now at step number 6 of our chart called "21-Step Integration of Tools: Manufacturing Process." We need to assess the process to determine the process capability/performance from the needs of the customer, as described by

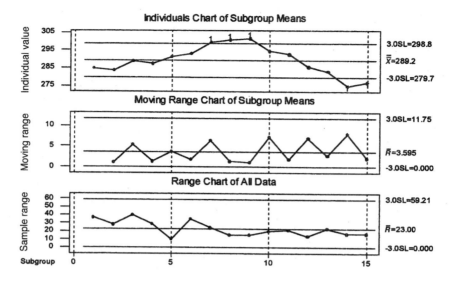

Figure 11.3 Three-way control chart

their specifications. If the process is considered not to be capable, we can then proceed on a process improvement effort, beginning with step 7 of the 21-step chart.

11.4 SHORT RUNS

A control chart is often thought to be a technique for controlling the characteristics or dimensions of products, the thought being that a controlled process would yield products that are more consistent. However, the dimension of a part, for example, is the end result of a process.

Typically, it is more useful to focus on key product characteristics and their related process parameters rather than on a single product characteristic. Brainstorming techniques can help with this selection activity. It is most effective to make use of the minimum number of charts while at the same time maximizing the usefulness of the charts.

Even so, it is tempting to think that control chart techniques are *not* useful in a given situation for one or more of the following reasons:

- Insufficient parts in a single production run
- Small lot sizes of many different parts
- Completion time is too short for the collection and analysis of data, even though the production size is large

Note that the emphasis in these three application scenarios is product measurement. Our thinking about the applicability of control charting to these situations changes when we use a methodology that links these product scenarios to process measurements. Chapter 34 of Breyfogle (1999), *Implementing Six Sigma*, describes short-run control charting, which makes a bridge from product to process measurements. Future part numbers from a process can benefit from the current wise application of control charting and process improvement activities.

Manufacturing and business process applications for short-run charts include

- Solder paste thickness for circuit boards
- Inside diameter of extrusion parts
- Cycle time for purchase orders
- Delivery time for parts

11.5   ENGINEERING PROCESS CONTROL

Consider a situation where parts are produced by a tool that needs periodic sharpening, adjustment, or replacement. These situations are application examples of exponentially weighted moving average (EWMA) control charts with engineering process control (EPC).

In the Shewhart model for control charting, the mean is assumed to be constant. Also, errors are assumed to be normal, independent with zero mean, and with a constant variance σ^2. In many applications, these assumptions do not hold true. EWMA techniques offer an alternative that is based on exponential smoothing, sometimes called geometric smoothing.

Taking the weighted average of past observations with progressively smaller weights over time gives the EWMA computation the effect of a filter. The addition of a weight factor to achieve balance between older data and more recent observations gives EWMA flexibility of computation.

EWMA techniques can be combined with engineering process control (EPC) to give insight into when a process should be adjusted. The mechanics of conducting an EWMA and EPC are described in Chapter 36 of Breyfogle (1999), *Implementing Six Sigma*.

11.6   EXAMPLE 11.3: OPTIMIZING NEW EQUIPMENT SETTINGS

The performance of a new soldering oven needed to be optimized. A 16-trial, two-level fractional factorial DOE with the following factors and factor levels were considered:

Factor	Factor Level	
	−	+
Zone 1	220	250
Zone 2	230	250
Zone 3	230	250
Belt Speed	7 in	9 in.
Bars/Trays	Bar	Tray
Plating	Matte tin	Bright tin
Cover/Shield	With cover	Without cover

The experiment's results led to a process optimization that cut the number of rejects in half as compared to the performance of the previous equipment. In addition, data analysis revealed that the pin protectors (i.e., cover/shield) were not needed. This permitted an employee to be moved to another process.

The mechanics of setting up and conducting fractional factorial DOEs are described in Chapters 27–33 of Breyfogle (1999), *Implementing Six Sigma*.

12

SERVICE/TRANSACTIONAL APPLICATIONS

Many companies have learned a key lesson in their implementation of Six Sigma: successful outcomes are not limited to manufacturing processes. A number of our clients have had great success applying S^4 methodology to their key service/transactional processes. Consider the following cases:

- *Customer 1 (Software Company)*: Implemented control charting and process capability/performance strategies in their customer order fulfillment process. Results: Subprocess owners in assembly, materials, planning, purchasing, shipping, and testing identified important measurements and process improvement opportunities using control charting, Pareto charting, normal probability plotting, and DOE tools.
- *Customer 2 (Pharmaceutical Testing Company)*: Created a strategy using SPC techniques to track iterations of paperwork. Result: Control charts and Pareto charts, in conjunction with brainstorming, could then indicate the magnitude of common cause variability and opportunities for improvement.
- *Customer 3 (Aerospace Service Company)*: Created a strategy to reduce the warranty return rate after servicing products. Result: Control chart techniques could monitor the proportion of returns over time, while Pareto charts could monitor the types of failures. Efforts could then be directed to improving processes on the "big hitter" items.
- *Customer 4 (City Government)*: Created a strategy to evaluate the differences between city home inspectors. Result: Random sampling and con-

trol charting techniques could quantify the differences over time. Through Pareto charting and brainstorming techniques, opportunities for improvement could be identified.

- *Customer 5 (School District):* Created a fractional factorial DOE strategy to assess the factors that could affect attendance and what could be done differently to bring about improvement.

These examples illustrate the many implementation strategies and unique tool combinations available for service/transactional projects. This chapter includes thoughts on what can be done when measuring service-type processes, an overall application strategy, and two examples of teams that have effectively utilized the S^4 methodology to improve key service/transactional processes.

12.1   MEASURING AND IMPROVING SERVICE/ TRANSACTIONAL PROCESSES

Arriving at a meaningful definition of defects and collecting insightful metrics are often the biggest challenges in attempts to apply the S^4 methodology to service/transactional processes. Projects of this type sometimes lack objective data. When data do exist, the practitioner is usually forced to work with attribute data such as pass/fail requirements or number of defects. Teams should strive for continuous data over attribute data whenever possible; continuous data provide more options for statistical tool usage and yield more information about the process for a given sample size. A rough rule of thumb is to consider data continuous if at least 10 different values occur and no more than 20% of the data set are repeat values.

Frequently in service/transactional projects, a process exists and a goal has been set, but there are no real specification limits. Setting a goal or soft target as a specification limit for the purpose of determining process capability/performance indices can yield questionable results. Some organizations spend a lot of time creating an arbitrary metric by adding a specification value where one is not appropriate. Consider the impact of forcing a specification limit where one does not exist. Then consider using probability plotting to gain insight into the "capability of a process" expressed in units of "frequency of occurrence." Chapter 11 of Breyfogle (1999), *Implementing Six Sigma*, gives more detailed information on planning, performing, and analyzing the results of process capability/ performance studies for this type of situation.

A measurement systems analysis (MSA) should be considered even for service/transactional processes. If attribute data are gathered, the importance of performing an MSA to determine the amount of variability the "Gage" is adding to the total process variation should not be overlooked. Attribute Gage studies can also be conducted by comparing how frequently measurements agree for the same appraiser (repeatability) and between appraisers (reproducibility).

Chapter 12 of Breyfogle (1999), *Implementing Six Sigma*, contains more detailed information on planning, performing, and analyzing the results of measurement system studies.

It requires persistence and creativity to define process metrics that give true insight into service/transactional processes. Initial projects may take longer to complete owing to up-front work needed to establish reliable measurement systems. However, many of the "low-hanging fruit" projects can be successfully attacked with cause-and-effect-matrices or employee surveys. These tools can help teams determine where to initially focus while establishing data collection systems to determine the root cause of the more difficult aspects of a project.

Lastly, DOE techniques are frequently associated with manufacturing processes, but they can also provide significant benefits to service/transactional projects as well. A well-designed DOE can help establish process parameters to improve a company's efficiency and quality of service. The techniques offer a structured, efficient approach to experimentation that can provide valuable process improvement information.

12.2   21-STEP INTEGRATION OF THE TOOLS: SERVICE/TRANSACTIONAL PROCESSES

There are many possible alternatives and sequencing patterns of Six Sigma tools for S^4 service/transactional projects. Our "21-Step Integration of the Tools: Service/Transactional Processes," shown below, is a concise summary of how Six Sigma tools can be linked or otherwise considered for service/transactional projects. This road map can be referenced for insight on how individual tools fit into the "big picture." Specific tools are highlighted in bold print to aid the reader in later locating where a tool may be applied. The glossary at the back of the book will help to clarify terms. Breyfogle (1999), *Implementing Six Sigma*, provides details on the mechanics of using the tools.

21-Step Integration of the Tools: Service/Transactional Processes

Step	Action	Participants	Source of Information
1	Identify critical customer requirements from a high-level project measurement point of view. Define the scope of projects. Identify **KPOVs** that will be used for project metrics. Establish a **balanced scorecard** for the process that considers also **COPQ** and **RTY** metrics.	Six Sigma practitioner and manager championing project	Organization wisdom

2	Identify team of key "stakeholders" for project. Address any project format and frequency of status reporting issues.	Six Sigma practitioner and manager championing project	Organization wisdom
3	Describe business impact. Address financial measurement issues of project.	Six Sigma practitioner and Finance	Organization wisdom
4	Plan overall project. Consider using this "21-Step Integration of the Tools" to help with the creation of a project management **Gantt chart**.	Team and manager championing project	Organization wisdom
5	Start compiling project metrics in time series format with sampling frequency reflecting "long-term" variability. Create **run charts** and **"30,000-foot level" control charts** of KPOVs. Control charts at this level can reduce amount of "firefighting."	Six Sigma practitioner and team	Current and collected data
6	Since specifications are not typically appropriate for transactional processes, describe "long-term" process **"capability/performance"** of KPOVs in "frequency of occurrence" units. For example, 80% of the time a KPOV response is within a particular range. Describe implication in monetary terms. Use this value as baseline performance. **Pareto chart** types of nonconformance issues.	Six Sigma practitioner and team	Current and collected data
7	Create a **process flowchart/process map** of the current process at a level of detail that can give insight to what should be done differently.	Six Sigma practitioner and team	Organization wisdom
8	Create a **cause-and-effect diagram** to identify variables that can affect the process output. Use the process flowchart and Pareto chart of nonconformance issues to help with the identification of entries within the diagram.	Six Sigma practitioner and team	Organization wisdom

9	Create a **cause-and-effect matrix** assessing strength of relationships between input variables and KPOVs. Input variables for this matrix could have been identified initially through a cause-and-effect diagram.	Six Sigma practitioner and team — Organization wisdom
10	Even transactional/service processes should consider the benefit of conducting a **measurement systems analysis**. This study may take the form of an attribute measurement systems analysis study. Within transactional processes people, for example, may not be classifying nonconformance similarly.	Six Sigma practitioner and team — Active experimentation
11	Rank importance of input variables from the **cause-and-effect matrix** using a **Pareto chart**. From this ranking create a list of variables that are thought to be KPIVs.	Six Sigma practitioner and team — Organization wisdom
12	Prepare a focused **FMEA**. Consider creating the FMEA from a systems point of view, where failure mode items are the largest–ranked values from a **cause-and-effect matrix**. Assess current control plans.	Six Sigma practitioner and team — Organization wisdom
13	Collect data for assessing the KPIV/KPOV relationships that are thought to exist.	Six Sigma practitioner and team — Collected data
14	Use **multi-vari charts, box plots**, and other graphical tools to get a visual representation of the source of variability and differences within the process.	Six Sigma practitioner and team — Passive data analysis
15	Assess statistical significance of relationships using **hypothesis tests**.	Six Sigma practitioner and team — Passive data analysis
16	Consider using **variance components analysis** to gain insight to the source of output variability. Example sources of variability are day-to-day and department-to-department differences.	Six Sigma practitioner and team — Passive data analysis

17 Conduct **correlation, regression**, and **analysis of variance** studies to gain insight into how KPIVs can impact KPOVs.	Six Sigma practitioner and team	Passive data analysis
18 Consider using fractional factorial **DOEs** to assess the impact of process change considerations within a process. This assessment approach can give insight to interactions that may exist, for example, between noise factors and change considerations.	Six Sigma practitioner and team	Active experimentation
19 Determine optimum operating windows of KPIVs from **DOEs** and other tools.	Six Sigma practitioner and team	Passive data analysis and active experimentation
20 Update **control plan**. Implement "50-foot level" **control charts** to timely identify special cause excursions of KPIVs.	Six Sigma practitioner and team	Passive data analysis and active experimentation
21 Verify process improvements, stability, and **capability/ performance** using demonstration runs. Create a final project report stating the benefits of the project, including bottom-line benefits. Make the project report available to others within the organization. Monitor results at 3 and 6 months after project completion to ensure that project improvements/benefits are maintained.	Six Sigma practitioner and team	Active experimentation and/or collected data

12.3   EXAMPLE 12.1: IMPROVING ON-TIME DELIVERY

As part of its strategic business goals, a company gathered a cross-functional group of employees, including those who knew the details of the process, and appointed an S^4 improvement team to increase on-time delivery from their suppliers to their 20 service centers.

The team decided on a primary project metric: percent of items late (attribute data). Samples from the service centers were taken at a rate of 100 purchase orders a week for one month in order to create a "30,000-foot level" view of the process. Table 12.1 shows the data sampled for creation of the control chart. Note that the control chart limits are a function of sample size and that variability between samples is not taken into account when calculating the limits. The data were collected on a late/not-late criterion and were therefore tracked using a p chart. As Figure 12.1 shows, the process is in statistical control but has a presumably unacceptable overall late-delivery rate of approximately 61% (Chapter 10 of Breyfogle [1999], *Implementing Six Sigma*, discusses other control charting possibilities).

TABLE 12.1   Data for Late Deliveries Example

Sample Number	Number of Late Items	Nonconforming Fraction
1	48	.48
2	68	.68
3	71	.71
4	49	.49
5	67	.67
6	69	.69
7	69	.69
8	61	.61
9	70	.70
10	62	.62
11	65	.65
12	55	.55
13	63	.63
14	69	.69
15	51	.51
16	62	.62
17	71	.71
18	50	.50
19	47	.47
20	60	.60

The Black Belt compiled a database of all pertinent details on the sampled purchase orders. She analyzed the data, looking for trends in late deliveries. Figure 12.2 shows a Pareto chart of the data by service center. She facilitated the team in a brainstorming session to compile a list of possible causes for late deliveries within a cause-and-effect diagram. During the meeting, service center managers denied having such a drastic late-delivery problem, but the data proved them wrong. Headquarters thought the service center managers were

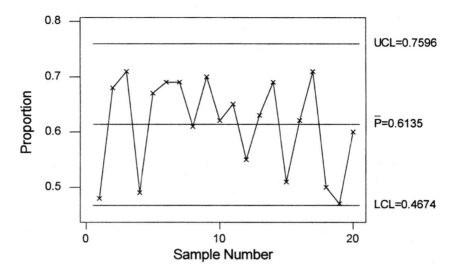

Figure 12.1   *p* chart example for late-delivery data

not motivated to change. Team members thought the project was a waste of time. Overall buy-in was low. The Six Sigma practitioner expended a lot of time looking for trends by service center, suppliers, materials, and time of year, but no trends that would account for late deliveries were immediately apparent.

The Black Belt decided to create a detailed map of the purchasing process. By mapping the purchase-order entry process in detail, the team determined that under certain circumstances material request dates were defaulting to the same day the order was placed. Owing to incomplete purchase-order forms, more than 60% of the late purchase orders were placed in the emergency order category—that is, orders needed the same day. Before a project could be undertaken that truly improved on-time customer delivery, improvements to the purchase-order entry module were needed.

The team decided to mistake-proof the process and made the material request date a required field. If same-day delivery was entered on the order form, the order-entry person was prompted to select a reason for the emergency order. Through changes in the computer system and training of order-entry personnel, the process went from 1.02 sigma to 2.71 sigma. The before and after outputs to the process that illustrate this dramatic improvement are shown in the Pareto chart in Figure 12.3.

At this point charts of the new data, similar to Figure 12.4, can be re-created. Any points beyond the control limits should be investigated. The process de-

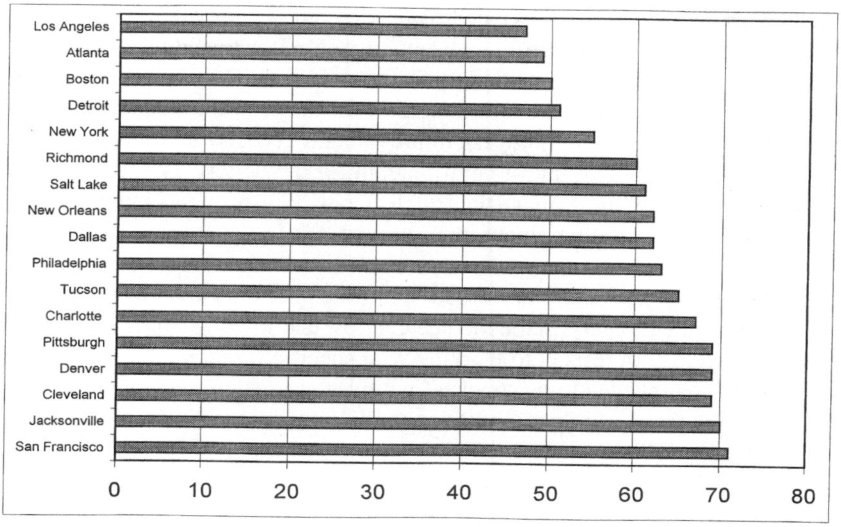

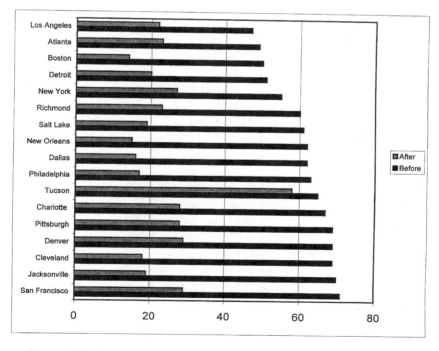

Figure 12.2   Pareto chart of original late-delivery data

Figure 12.3   Pareto chart of late-delivery data before and after process change

scribed above can then be repeated using the "good" data, and a root-cause analysis of late deliveries can again be expressed in Pareto chart format to provide direction for further process improvement efforts. Control chart tracking can then be used in greater detail to monitor the largest Pareto chart items and provide insight as to when process improvements focused on these items are effective (i.e., should go out of control again for the better). The team should also translate any of these quantifiable defect-rate improvements into monetary benefits for presentation to upper management.

While it proved difficult to calculate "hard" monetary savings to this initial project, altering the order-entry process was the necessary foundation for improving on-time delivery in the service centers, and it led to improvements in the bottom line with respect to working with specific suppliers. It was the first "big hitter" that needed to be attacked, but this was difficult to see in the data. Now that the team had "good" data to work with, objectivity and buy-in increased.

It should be noted that for this project the data could have been transformed from attribute (i.e., order was late) to continuous (i.e., number of days late). The response charting criterion could be changed from pass/fail to number of days late. A team could then create a "30,000-foot level" control chart, randomly selecting one order per day, to obtain a long-term view of the process output.

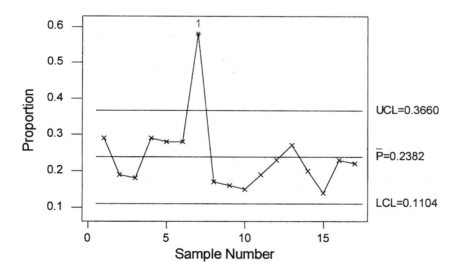

Figure 12.4   *p* chart showing new late-delivery data

12.4   EXAMPLE 12.2: APPLYING DOE TO INCREASE WEBSITE TRAFFIC

A company was unhappy with the number of "hits" its website had received and the website's rank with search engines. The company formed an S^4 improvement team to increase traffic on the website. Upper management gave the team a budget and an aggressive goal: doubling the number of hits within the next four months.

The team compiled "hit-rate" data in time series format from the website, looking at hit rates by week over the past three months. The data used for the project are shown in Table 12.2, and an *XmR* control chart of this data is shown in Figure 12.5. Note that weeks 4 and 5 are beyond the control limits. Investigation of these two data points showed that they reflected website traffic during the last two weeks of December, when all websites generally have lower activity owing to the holiday season. Since these data points could be accounted for, they were removed from the data and the control charts were re-created.

TABLE 12.2   Website "Hit-Rate" Data

Week	Number of Hits
11/7/99	4185
11/14/99	3962
11/21/99	3334
11/28/99	4176
12/5/99	3889
12/12/99	3970
12/19/99	2591
12/26/99	2253
1/2/00	4053
1/9/00	5357
1/16/00	5305
1/23/00	4887
1/30/00	5200
2/6/00	4390
2/13/00	4675
2/20/00	4736
2/27/00	4993

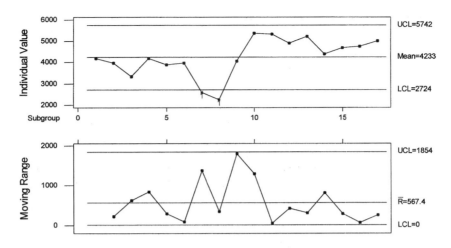

Figure 12.5 XmR control chart of website hit rates

Next, the team held a brainstorming session to create a cause-and-effect diagram. They came up with many factors that might increase the number of page requests on the website. To rank this list of factors and begin to determine the KPIVs, the team created a cause-and-effect matrix similar to Table 12.3.

TABLE 12.3 Cause-and-Effect Matrix: Factors That Affect Website Traffic

Process Input	Increase Website Hit Rate	Easy to Implement	◄ Process Outputs
	10	10	◄ Importance
	Correlation of Input to Output		Total
Number of keywords	9	9	180
Type of keywords	9	9	180
Length of description	9	9	180
URL	3	9	120
Frequency of updates	9	3	120
Description	3	9	120
Free gift	9	1	100
Search engine used	3	3	60
Contest	3	1	40
Text size and font	1	3	40
Links from other sites	3	1	40
Banner advertising	1	1	20

No data were available on the relationships of the KPIVs to the website hit rate, so the team decided to run an experiment. To test the critical factors, they established different gateway pages with links back to the main site. Instead of testing these factors one at a time, the Six Sigma practitioner opted for a DOE approach. He knew that a DOE approach would make the team more efficient and allow them to gain a better understanding of how the critical factors interacted with each other and affected website hit rate.

To reduce the size of the experiment and the number of gateway pages needed, the team reviewed the list of factors from the cause-and-effect matrix and pared it down to factors that would yield insight specifically into website traffic. Next, a meeting of peers from all affected areas was conducted to consider expansion or deletion of selected factors, outputs, and limiting assumptions in the experiment. The team then grouped the factors into two levels, chosen to represent the extremes of each setting. The revised list, with settings, is shown in Table 12.4.

TABLE 12.4   Factors and Levels of Website Hit Rate Test

Factor Designation	+ Setting	− Setting
A	URL title, new	URL title, old
B	Length of descriptions: long	Length of description: short
C	Number of keywords: many (10 words)	Number of keywords: short (2 words)
D	Types of key words: new	Types of key words: old
E	Site updates: weekly	Site updates: monthly
F	Location of key words in title: 40th character	Location of key words in title: 75th character
G	Free gift: yes	Free gift: no

The Six Sigma practitioner used the procedure listed in Chapter 42 and Table M2 of Breyfogle (1999), *Implementing Six Sigma*, and came up with the test matrix shown in Table 12.5. Eight gateway pages were created, each page representing one of the trials listed in this table. Hit rates were then collected on a weekly basis to determine which combination of factors produced the most website traffic.

TABLE 12.5   DOE Test Matrix

Trial Number	A	B	C	D	E	F	G
1	+	−	−	+	−	+	+
2	+	+	−	−	+	−	+
3	+	+	+	−	−	+	−
4	−	+	+	+	−	−	+
5	+	−	+	+	+	−	−
6	−	+	−	+	+	+	−
7	−	−	+	−	+	+	+
8	−	−	−	−	−	−	−

Based on the results of the experiment, the team could determine which factors to focus on in its attempt to increase traffic on the company website. Using the results they could, for example, test the hypotheses developed by Web Consulting Honolulu concerning the most important things a company can do to increase website traffic (Web Consulting Honolulu, 1999). The consulting firm believes that having the right key words and offering a "gift" (e.g., a discount or a download) are key contributors or KPIVs to increase website traffic.

As part of the control plan, the team may choose to check the ranking of the website against those of other sites weekly, using a combination of keywords they have deemed important to the increasing interest in the site. Whenever the site ranking degrades, adjustments are considered and/or another DOE is conducted to address any new technology issues that may have a bearing on website traffic.

12.5   OTHER EXAMPLES

There are many other services/transactional project examples included in Breyfogle (1999), *Implementing Six Sigma*. The following list includes some of the more notable ones:

- Process measurement and improvement within a service organization: Example 5.3.
- A 20:1 return on investment (ROI) study could lead to a methodology that has much larger benefit through the elimination of the justification process for many capital equipment expenditures: Example 5.2.
- A described technique better quantifies the output, including variability, of a business process that has no specifications: Example 11.5.

- The bottom line can be improved through the customer invoicing process: Example 43.8.
- An improvement in the tracking metric for change order times can give direction to more effective process change needs: Example 43.6.
- Illustrating how S^4 techniques improve the number in attendance at a local ASQ section meeting (the same methodology would apply to many process measurement and improvement strategies): Example 11.5 and Example 19.5.

13

DEVELOPMENT APPLICATIONS

Six Sigma techniques are applicable to a wide variety of situations, including development. The application of Six Sigma techniques to the development process is often called Designed for Six Sigma (DFSS). The following examples illustrate how Smarter Solutions has helped its clients benefit from the application of Six Sigma tools.

- *Customer 1 (Hardware/Software Integration Test Organization):* Used a unique test strategy for evaluating the interfaces of new computer designs. Result: The methodology was more effective and efficient in determining if a computer would have functional problems when interfacing with various customer hardware and software configurations.
- *Customer 2 (Product Mixture Development Organization):* Used a mixed DOE design to develop a solder paste. Results: The analysis identified significant mixture components and opportunities for improvement.
- *Customer 3 (Reliability Test Organization):* Developed a unique reliability test DOE strategy to evaluate preship product designs. Result: Monetary savings from the detection of combination manufacturing and design issues before first customer shipment.
- *Customer 4 (Printer Design Organization):* Used DOE techniques to assess the validity of a design change that was to reduce the settle-out time of a stepper motor. Result: The DOE validated the design change and the amount of improvement. In addition, the DOE indicated the significance of another adjustment.

- *Customer 5 (Computer Company):* Created a unique metric feedback strategy within a system test. Result: Improved feedback loops when testing a product, between product evaluations, and field-to-design failure mechanisms would significantly decrease expenses through failure prevention.
- *Customer 6 (Government Certification Test Organization):* Improved electromagnetic interference (EMI) testing. Result: An efficient approach was developed that considers both within- and between-machine variability.
- *Customer 7 (Hardware Design Test Organization):* Created an improved reliability test strategy for a complex power supply. Result: A less expensive DOE-based reliability strategy identified exposures that a traditional reliability test would not capture.
- *Customer 8 (Reliability Organization):* Used DOE techniques in a stress test environment to improve design reliability. Result: The component life was projected to increase by a factor of 100.

These examples illustrate the many implementation strategies and unique tool combinations available for development processes. This chapter includes thoughts on what can be done when measuring development processes, an overall application strategy, and application examples.

13.1 MEASURING AND IMPROVING DEVELOPMENT PROCESSES

In DFSS contexts we suggest the application, whenever possible, of process measurements to the various products that go through the development process. Product-to-product differences can then be viewed as noise to the development process measurements. Even where this measurement technique is not appropriate, the techniques described within this chapter are still applicable.

Suppose, for example, that the functional test area for new computer designs is to be measured and then improved. This situation is often not viewed as a process where the output of the process crosses product boundaries. Typically, the focus is on the created product, not the process that develops the product. Information is often not compiled to quantify the effectiveness of current methodologies in creating new products that are developed timely to meet manufacturing needs along with the quality and functional needs of the end user. Much can be gained if this way of thinking is altered in such a way that a feedback and measurement system is created, as illustrated in Figure 13.1, where the chronicle sequence unit within the control chart is products that were developed over time.

Most Six Sigma tools are applicable in a DFSS context; however, strategies for and the sequence of use of these tools can be very different from what might

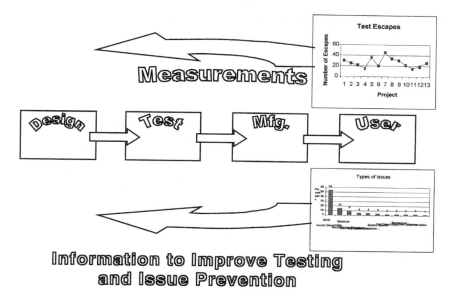

Figure 13.1   Measurements across products developed over time with information feed-back from manufacturing to development

be expected within manufacturing or service/transactional processes. For example, fractional factorial DOEs are often used to fix problems in manufacturing; however, in development, structured fractional factorial DOEs could be used as a control measure for achieving a high-quality product design. A measure for the effectiveness of a design FMEA within a development process can be considered a control measure for creating a product that consistently meets or exceeds the needs of customers.

13.2   21-STEP INTEGRATION OF THE TOOLS: DEVELOPMENT PROCESSES

There are many possible alternatives and sequencing patterns of Six Sigma tools for S⁴ development projects. Our "21-Step Integration of the Tools: Development Processes," which follows, is a concise summary of how Six Sigma tools can be be linked or otherwise considered for development projects. This road map can be referenced to give insight as to how individual tools fit into the "big picture." Specific tools are highlighted in bold print to aid the reader in later locating where a tool may be applied. The glossary in the back of the book can help clarify terms. Breyfogle (1999), in *Implementing Six Sigma*, provides details on the mechanics of tool utilization.

21-Step Integration of the Tools: Development Processes

Step	Action	Participants	Source of Information
1	**Brainstorm** for linkages to previous products that were developed for the purpose of creating a set of measurements that can bridge from the development of a product to the process of developing products.	Six Sigma practitioner and manager championing project	Organization wisdom
2	Use **QFD** when appropriate to define customer requirements. Identify KPOVs that will be used for project metrics. Establish a **balanced scorecard** for the development process that considers also **COPQ** and **RTY** metrics for the development process.	Six Sigma practitioner, manager championing project, and customer	Organization wisdom
3	Identify team of key "stakeholders" for projects that are to improve the development process. Define the scope of the projects. Address any project format and frequency of status reporting issues.	Six Sigma practitioner and manager championing project	Organization wisdom
4	Describe business impact. Address financial measurement issues of project. Address how cost avoidance will be credited to the project financials. Plan a methodology for estimating the future savings from the project. Consider using the results from **DOE** analyses to make a prediction for financial project benefits.	Six Sigma practitioner, manager championing project, and Finance	Organization wisdom
5	Plan overall project. Consider using this "21-Step Integration of the Tools" to help with the creation of a project management **Gantt chart**.	Six Sigma practitioner, manager championing project, and team	Organization wisdom

6	When possible, compile metrics from products that went through a similar development cycle. Present these metrics in time series format where each time unit is a chronical sequencing of previous products. Create **run charts** and "30,000-foot level" **control charts** of KPOVs such as development cycle time, number of customer calls/problems for the product, and ROI of the product. Control charts at the "30,000-foot level" can show when the responses from some products from the development process were "different" from other developed products from a special cause(s). When products have responses within the control limits, the differences between the products could be considered as "noise" to the overall development process.	Six Sigma practitioner and team	Historical data collection
7	When measuring the development process, specifications are not typically appropriate. "Long-term" process "capability/performance" of KPOVs can then be described in "frequency of occurrence" units. For example, 80% of the time the development cycle time of a product is between 6.1 and 9.2 months. Another example: 80% of the time the number of reported problems for a developed product is between 2.8 and	Six Sigma practitioner and team	Current and collected data

4.3 calls per unit sold. Describe implication in monetary terms. Use this value as a baseline for the project. **Pareto chart** types of nonconformance issues from previous products that were developed. This Pareto chart might indicate that responses to customer questions about product setup had more of a financial impact to the business than did product reliability service issues.

8	Create a **process flowchart/ process map** of the current development process at a level of detail that can give insight to what should be done differently.	Six Sigma practitioner and team — Organization wisdom
9	Create a **cause-and-effect diagram** to identify variables that can affect the process output. Use the **process flowchart** and **Pareto chart** of nonconformance issues to help with the identification of entries within the diagram.	Six Sigma practitioner and team — Organization wisdom
10	Create a **cause-and-effect matrix** assessing strength of relationships between input variables and KPOVs. Input variables for this matrix could have been identified initially through a **cause-and-effect diagram**.	Six Sigma practitioner and team — Organization wisdom
11	Rank importance of input variables from the **cause-and-effect matrix** using a **Pareto chart**. From this ranking create a list of variables that are thought to be KPIVs.	Six Sigma practitioner and team — Organization wisdom
12	Prepare a focused design **FMEA** where the failure modes are components of the design. Consider also creating the FMEA from a systems point of view, where failure	Six Sigma practitioner and team — Organization wisdom

mode items are the largest-
ranked values from a **cause-
and-effect matrix**. Assess
current control plans. Within
FMEA consider risks of
current development tests and
measurement systems.
Conduct **Measurement
Systems Analyses** where
appropriate.

13	Collect data from previous development projects to assess the KPIV/KPOV relationships that are thought to exist.	Six Sigma practitioner and team	Collected data
14	When enough data exist from previously developed products, use **multi-vari charts**, **box plots**, and other graphical tools to get a visual representation of the source of variability and differences within the process.	Six Sigma practitioner and team	Passive data analysis
15	When enough data exist from previously developed products, assess statistical significance of relationships using **hypothesis tests**.	Six Sigma practitioner and team	Passive data analysis
16	When enough data exist from previously developed products, consider using **variance components analysis** to gain insight into the source of output variability. Example sources of variability are supplier-to-supplier and plant-to-plant variability.	Six Sigma practitioner and team	Passive data analysis
17	When enough data exist from previously developed products, conduct **correlation**, **regression**, and **analysis of variance** studies to gain insight on how KPIVs can impact KPOVs.	Six Sigma practitioner and team	Passive data analysis
18	Conduct fractional factorial **DOEs**, **pass/fail functional testing**, and **response surface analyses as needed to** assess KPOVs, considering factors	Six Sigma practitioner and team	Active experimentation

that are design, manufacturing, and supplier in nature. Consider structuring experiments so that the levels of KPIVs are assessed relative to the reduction of variability in KPOVs when the product will later be produced within manufacturing. Consider structuring the experiment for the purpose of determining manufacturing KPIV settings that will make the process more robust to noise variables such as raw material variability. Consider how wisely applied DOE techniques can replace and/or complement existing preproduction **reliability tests** that have been questionable in predicting future product failure rates. DOE within development could be used and considered a Six Sigma control methodology for the development process.

19 Determine from **DOE**s and other tools optimum operating windows of KPIVs that should be used within manufacturing.	Six Sigma practitioner and team	Passive data analysis and active experimentation
20 Update **control plan,** noting that fractional factorial DOE experiments, pass/fail functional testing methodologies, and wisely applied reliability assessments can be an integral part of the control plan of the development process. Also, consider within this control plan the involvement of development and other organizations whenever field problems occur for the purpose of immediate problem resolution and improvements to the overall development process for future products.	Six Sigma practitioner and team	Passive data analysis and active experimentation

| 21 | Verify process improvements, stability, and **capability/ performance** using demonstration runs. Create a final project report stating the benefits of the project, including bottom-line benefits. Make the project report available to others within the organization. Consider monitoring the execution and results closely for the next couple of projects after project completion to ensure that project improvements/benefits are maintained. | Six Sigma practitioner and team | Active experimentation |

13.3 EXAMPLE 13.1: NOTEBOOK COMPUTER DEVELOPMENT

Notebook computers are developed and manufactured by a company. Product reliability is important to the company, since product problems are expensive to resolve and can impact future sales. One KPOV measure for this type of product was the number of machine problem phone calls and/or failures per machine over its product life.

The frequency of this metric was tracked using a "30,000-foot level" control chart, where the chronological sequencing of the chart was type of product. The frequency of this KPOV was judged to be too large. An S^4 project was defined to reduce the magnitude of this number.

As part of the S^4 project, a Pareto chart was created. This chart indicated that the source of most customer problems was "customer setup" followed by "no trouble found." A "no trouble found" condition exists whenever a customer states that a problem existed but the failure mode could not be re-created when the machine was returned to a company service center.

To reduce the number of customer setup problems, a team was created to identify the process used to establish customer setup instructions. A cause-and-effect diagram was then created to solicit ideas on what might be done to improve the process for the purpose of reducing the amount of confusion a customer might have when setting up a newly purchased machine. A cause-and-effect matrix was then created to rank the importance of the items with the cause-and-effect diagram. Some "low-hanging fruit" was identified that could be addressed immediately to improve the process of creating customer setup instructions. An FMEA was then used to establish controls to the highest-ranked items from the cause-and-effect matrix. As part of the improvement effort and control mechanism for developing new instruction procedures, a DOE matrix would be used to conduct the testing of new instructions; the factors from this matrix would include items such as experience of the person setting up the new computer design.

For the second-highest category, "no trouble found," in the Pareto chart a cause-and-effect diagram, a cause-and-effect matrix, and then FMEA also indicated other control measures that should be in place. Examples of these types of control measures—found in Breyfogle (1999), *Implementing Six Sigma*—that could be identified through S^4 projects are:

- Testing for thermal design problems that are causing NTF: Example 30.2
- Creating a more effective test for reliability issues using fewer resources through DOEs: Example 43.2
- Improving software/hardware interface testing: Example 42.4
- Managing product development of software: Example 42.6
- Tracking of change orders: Example 43.6
- Creating a more effective ORT test strategy: Example 40.7
- Using a system DOE stress-to-fail test to quantify design margins: Example 31.5
- Using a DOE strategy to size the future capability of a product by testing a small number of preproduction samples. Methodology can also determine if any key process manufacturing parameters should be monitored or have a tightened tolerance: Example 30.1
- Using a QFD in conjunction with a DOE to determine manufacturing settings: Example 43.2

14

NEED FOR CREATIVITY, INVENTION, AND INNOVATION

For more than a decade *Fortune* magazine has been ranking major public U.S. companies on "innovation." Their findings are insightful (Jonash and Sommerlatte, 1999). The top 20% of firms rated high on innovation also have twice the shareholder rate of return as compared to other firms in their industry. The bottom 20%, on the other hand, report shareholder returns 33% below those of their industrial peers. These data suggest that innovation likely plays a considerable role in corporate profitability and shareholder wealth.

Six Sigma needs to be more than metrics and improvements to existing processes. To maintain customer relationships, a company must create a culture in which innovation occurs continuously within the company's services and products. The needs of the customer are dynamic. Product features that were considered "wow" features in the past are now taken for granted. For example, a person seeing for the first time a car that does not need a "crank start" would consider this a "wow" change; however, electronic starters are now taken for granted. Noritaki Kano's description of this concept is illustrated in Figure 14.1 (King, 1987).

The diagonal arrow in the middle, "one-dimensional quality," shows the situation where the customers tell the producer what they want and the producer supplies this need. The lower arrow represents the items that are expected; customers are less likely to mention these, but they are dissatisfied if they do not receive them. Safety is an example of items in this category. The top arrow represents "wow" quality; these are the items that a normal customer will not mention. Innovative individuals within the producer's organization or "leading edge" users need to anticipate these items, which must be satisfiers, not dissatisfiers.

A customer's "hidden" needs are not even on the customer's radar screen of desirable properties. Many times these unexpected attributes are invisible to customers because their belief system says they are technically unfeasible. But there are techniques that can tease these latent wants and needs out into the open. We need to design and integrate highly desirable attributes and features into an already complex, sometimes conflicting, assortment of customer requirements.

We suggest that the measurement category "innovation and learning" be considered when creating a balanced scorecard. Balanced scorecard metrics should be tied to the organization's strategic objectives. Any organization that hopes to remain globally competitive may instead have a short life if it ignores the need for continuous, exemplary learning, creativity, and innovation.

This chapter gives thoughts on how to create the culture, infrastructure, and test bed of structured chaos that leads to the creation of new, wonderful, unexpected, and unimagined goods and services. There are many books and journal articles on these three topics. We will only glance at the tip of the iceberg in this brief overview. For those interested in further reading on this important topic, we have included an extensive reading listing in Table 14.1. Our top ten recommendations are highlighted in the table.

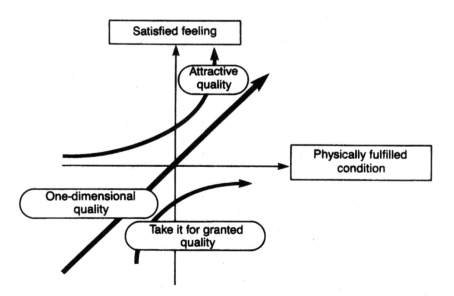

Figure 14.1   Satisfied feeling versus condition

TABLE 14.1   Recommended Reading List: Creativity, Invention, and Innovation

---

Top 10 in Date Order

SERIOUS PLAY, Michael Schrage (2000)

THE KNOWLEDGE-CREATING COMPANY, I. Nonaka and H. Takeuchi (1995)

DEFYING THE CROWD, Robert J. Sternbert and Todd I. Lubart (1995)

CREATIVITY, Robert W. Weisberg (1993)

INVENTION: THE CARE AND FEEDING OF IDEAS, Norbert Wiener (1993)

SERIOUS CREATIVITY, Edward De Bono (1992)

INVENTIVITY, John J. Gilman (1992)

THE FIFTH DISCIPLINE, Peter Senge (1990).

NOVATIONS, Gene W. Dalton and Paul H. Thompson (1986)

INNOVATION AND ENTREPRENEURSHIP, Peter F. Drucker (1985)

Next 20 in Date Order

CORPORATE CREATIVITY, Alan Robinson & Sam Stern (1998)

CRACKING CREATIVITY, Michael Michalko (1998)

UNLEASHING THE KILLER APP, Larry Downes U Chunka Mui (1998)

THE INNOVATOR'S DILEMMA, Clayton Christensen (1997)

ORGANIZING GENIUS, Warren Bennis and Patricia Ward Biederman (1997)

CREATIVITY: Mihaly Csikszentmihalyi (1996)

WELLSPRINGS OF KNOWLEDGE, Dorothy Leonard-Barton (1995)

DIFFUSION OF INNOVATIONS (4th Edition), Everett M. Rogers (1995)

WORKING WISDOM, Robert Aubrey and Paul M. Cohen (1995)

INNOVATE OR EVAPORATE, James M. Higgins (1995)

EMERGING PATTERNS OF INNOVATION, Fumio Kodoma (1995)

THIRD GENERATION R&D, P. A. Roussel, K. N. Saad, and T. J. Erickson (1991)

UNCOMMON GENIUS, Denise Shekerjian (1990)

FIRE IN THE CRUCIBLE, John Briggs (1988)

SCIENCE IN ACTION, Bruno Latour (1987)

OPPORTUNITIES, Edward De Bono (1978)

MANAGING THE FLOW OF TECHNOLOGY, Thomas J. Allen (1977)

SCIENTISTS IN ORGANIZATIONS, Donald Pelz and and Frank Andrews (1976)

THE ACT OF CREATION, Arthur Koestler (1964)

THE SOURCES OF INVENTION, J. Jewkes, D. Sawers, and R. Stillerman (1959)

14.1 DEFINITIONS AND DISTINCTIONS

Creativity

Creativity is the discovery or elicitation of something new, usually in the form of knowledge. Invention is the creation of new and useful things. An innovation is the embodiment of those "things" into prototypes with the potential for becoming profitable commercial goods and services. Gilman (1993) would add a fourth category: commercial activity. And if he were forced to measure each one of the four using only a single metric for each, he would measure (1) basic research (creativity) by "papers published," (2) applied research (invention) by "patents received," (3) development (innovation) by "prototypes/artifacts" created, and (4) commercial activity by "profits."

"The primary objective of industrial research is to generate profits"—not knowledge (Collier and Gee, 1973). However, there is no assurance that the act of creation or discovery will lead to an innovative product or service that will sell at a profit. The presumption that there is a strong relationship between basic research and commercial success or supremacy is simply wrong (Gomory, 1990). "New knowledge—especially new scientific knowledge—is not the most reliable or most predictable source of successful innovations" (Drucker, 1985a). Are these findings at odds with the inferences drawn from the first paragraph in this chapter? They are not. We began this chapter talking about innovation. The preceding comments in this paragraph talk about creativity, basic research, and scientific knowledge. This entire subsection deals with the topic of creativity. There are important differences among creativity, invention, and innovation, as we will describe. Subsequent sections will address the topics of invention and innovation.

Creativity consists of three components: expertise, imaginative thinking, and intrinsic motivation (Hughes et al., 1993). Expertise serves as the bedrock for creativity. A little knowledge provides the same stimulus to the mind that a small clump of "sourdough starter" adds to a damp lump of kneaded bread dough. Imaginative thinking is just another name for creativity. And intrinsic motivation can strongly promote and influence the creative acts of individuals. Stated bluntly, creative individuals are intelligent, imaginative, and motivated. If any one of these three traits is missing, creativity is not likely to flourish.

There is a difference between one's creative ability and the innovativeness of one's output (Pelz and Andrews, 1976). Creativity is a personal trait defined as the ability to envision associations between ideas and/or objects that others do not see. This creativity, however, is merely an *input* to the creative process. Creative *outputs* are products with varying degrees of innovativeness and productiveness. Innovative outputs create new possibilities. Productive outputs yield advances along an already existing path. Innovation equates to Juran's "breakthrough" thinking. Productivity equates to continuous improvement, which is sometimes referred to as kaizen.

Many industrial firms avoid the discovery/creativity phase of invention, because during this phase the uncertainty surrounding the potential market success and commercial value of a new idea is so great as to make an investment seem very risky (Wiener, 1993). The uncertainties associated with research efforts projected even a couple of years into the future are so great as to make traditional financial calculations of return on investment meaningless, if not harmful (Roussel, 1991). Knowledge-based innovation also has the longest lead time of the seven sources of innovation identified by Drucker (1985a). The lead time for new knowledge to mature into artifacts and then to marketable products ranges from 25 to 30 years. Thomas Kuhn (1970) found that it takes about 30 years for scientists to displace an existing but incorrect scientific theory and embrace a new one. Is it any wonder, then, that "most companies—especially smaller ones with sales of less than, say, $1 billion—should probably not invest significant sums in fundamental research" (Roussel 1991)? An organization has to have extremely deep pockets to be able to survive the lengthy period required to grow an idea into a profit center. It is also conceivable that the financial consequences of consistent, long-term, continuous improvement efforts provide a higher rate of return than do breakthrough technologies (Dertouzos, 1990).

Those organizations willing to invest in creating new knowledge, with the hope of developing it into profitable new products and processes, generally centralize their research under the corporate umbrella to protect the R&D staff from day-to-day operational concerns, budget cuts, and so forth. Those organizations more interested in investing their money in incremental improvement activities of invention and innovation are generally situated closer to the profit center of the business and its customers.

There is a connection we can make between creativity and Six Sigma. Six Sigma improvements are more typically used to address customer concerns and desires relating to existing products. Six Sigma is a team-based approach to problem solving and process improvement. Creativity (knowledge creation), on the other hand, is often a solo event. Truly new ideas come one at a time (Stookey, 1980). Research teams certainly have a role to play in government and business, but that place is during the latter stages of innovation. Six Sigma can aid with creativity and knowledge creation by fashioning a focus on the support activities that empower creativity and basic research. Activities such as ordering supplies, ready access to computation, library services, equipment repair, machine shops, attendance at conferences, equipment availability, continuous education and training, qualified technicians, an adequate budget, a pleasant work environment, and so forth can be streamlined through Six Sigma. The reason for minimizing delays in providing these kinds of goods and services is not the convenience of the researcher. Rather, the goal is to expedite work and reduce cost (Gilman, 1992). Another way of looking at the potential link between Six Sigma and creativity is that Six Sigma can improve the overall process of nurturing creativity.

The measurement of creativity within R&D in terms of the balanced scorecard as defined by Kaplan and Norton (1996) could lead to the list below. In this case we are using an R&D example, but the same principles apply to any creative function—marketing, stock analysis, and so forth.

- *Financial*: Total annual person-hours of free research time contributed to company-related projects by graduate students due to professional staff members acting as thesis advisers.
- *Customer:* Total annual contact person-hours the professional staff spends visiting current or potential customer sites soliciting ideas from their marketing, research, development, test, quality, and manufacturing staffs.
- *Internal:* Average number of hours each professional staff member spends per pay period mentoring junior professionals, observing/measuring an internal company process, and so forth.
- *Innovation and Learning:* Average number of work hours per pay period each professional staff member spends learning a new discipline and/or keeping current through reading in their area of expertise. The target goal is four hours per week per professional.

Invention

One measure of whether or not someone has invented something is whether or not it can be patented. To qualify for a patent an invention must be "new, useful, and non-obvious" (Lubar, 1990). Sternberg and Lubart (1995) use the terms "novel and appropriate." The National Science Foundation (NSF) prefers the term "applied research" to the term "invention," but the terms are synonymous. For several decades, the NSF (1989) has stated that the aim of applied research is "gaining the knowledge or understanding [necessary] to meet a specific, recognized need." By contrast, the NSF states that "the objective of basic research is to gain more comprehensive knowledge or understanding of the subject under study, without specific applications in mind."

Comparing these two definitions, we again see the dichotomy between creating knowledge and inventing things. Creativity is totally enamored with discovery and couldn't care less about the utility of what is learned. Artistic people create that they might have joy. Likewise, invention is enamored with creative prodding and probing but acquires a structural skeleton in response to "customer" needs—frequently a physical need. If we look ahead to the process of innovation, we find the skeleton of invention now covered by a tough layer of rawhide in response to the abrasive human clawing and clamoring for utility. This layer of skin tends to hide much of the underlying creativity that gave innovation its initial breath of life. The prototypes and artifacts of innovation must be useful or they are discarded as useless.

Managing invention is incredibly difficult (Buderi, 1993). Success on complex tasks still depends on the talents of a few key individuals. Identifying those key individuals is also difficult, because inventors tend to excel in many disciplines. Their cross-functional knowledge seems to be the source of their unique ability to tackle difficult problems. Most organizations excel in those areas they value and measure. Inventive companies reward inventive behavior; hence patents. If an organization wants to be inventive, it needs to measure and reward inventive behavior.

Below is a sample "invention" balanced scorecard as defined by Kaplan and Norton (1996). This is a university faculty example—the Department of Business Administration in a College of Business. For the sake of brevity, we offer only one metric per category.

- *Financial:* Average weekly contact hours per faculty member spent with students. This metric needs to be normalized by an instructor's course load.
- *Customer:* The faculty's median, student course evaluation scores on the question "My instructor is an effective and accomplished educator" (1–10 scale). This median score will be compared with that of all other academic departments. A goal might be a ranking above the 90th percentile.
- *Internal:* Average annual number of published journal articles or invited conference presentations per faculty member addressing the topic of "improving the quality of higher education".
- *Innovation and Learning:* The 3-year running average percentage of the faculty that take sabbaticals working in for-profit businesses for at least six months.

We should stress again how difficult it is to find, mentor, and manage inventive personalities. "No great breakthroughs have ever been achieved by reasonable men" (Magaziner and Patinkin, 1989). In the example of a university professor described above, it may be impossible for "unreasonable men/women" to achieve tenure. It also takes an extremely accomplished leader to manage creative, inventive, innovative people.

Innovation

Innovation is the creation of useful artifacts. The NSF refers to artifacts as prototypes. These artifacts are initially too crude or too symbolic to be of immediate interest to an external customer. The innovations that flow from the R&D process do not necessarily result in commercial products. They are a necessary but insufficient condition for commercialization. A few artifacts flourish, while the majority do not survive.

The NSF defines development (innovation) as "the systematic use of the knowledge or understanding gained from [basic and applied] research directed toward the production of useful materials, devices, systems, or methods, including the design and development of prototypes and processes."

It is worth noting that effective scientists do not limit their involvement to either "pure science" or "pure development" activities, but rather maintain interest and involvement in both (Pelz and Andrews, 1976). They straddle a fence between creative freedom and functional demands. Innovation receives a portion of its nourishment at the breast of invention but leads a life of its own.

One useful way of thinking about overall creative performance is to divide it into two categories. One contributes to the scientific community in the form of knowledge, while the other contributes to the organization in the form of products and profits (Edwards and McCarrey, 1973). When innovation transitions to commercial success we have achieved the second, and preferred, act of creation: wealth. Drucker (1985a) defines innovation as "the act that endows resources with a new capacity to create wealth." Unfortunately, capacity does not always deliver on its promises.

Creativity refers to ideas, whereas innovation involves the translation of those ideas into tangible products and intangible services (Tatsuno, 1990). These products and services are not yet of commercial grade but are early visions of what the customer wants or needs. And unlike creativity, which tends to originate in the mind of the individual, innovation involves collaboration with others in a social context (Schrage, 1990). Innovation cannot take place as a virtuoso performance, because the complex demands of the customer require multidimensional inputs from a wide range of organizational players. A support team can help with the design, packaging, advertising, selling, and manufacturing.

Innovation has a twin sister, technology. The two terms may actually be synonymous for all practical purposes. Price (1990) states that "technology equals scientific knowledge applied to the affairs of business." The innovation phase of product development is where the idea of technology begins to surface. Prototypes and artifacts suggest the inventor or creator has preliminary thoughts and visions of what the tangible product or intangible service should "look like." The word *technology* is derived from the Greek word for art or skill. Wiener (1993) wisely observes that the process of invention "is not complete until it reaches the craftsman." These craftsmen are the builders of the artifacts and prototypes of innovation.

Innovation has another interesting dimension. Based on the evaluation of 154 scientists and engineers, the correlation between innovation and productivity was 0.735 (Stahl and Steger, 1977). Innovation is not found without productivity. And this brings us to the vital subject of simulations, prototypes, and rapid prototyping. The relationship between innovation, productivity, and prototyping (specifically, rapid prototyping) will soon become clear. The bulk of the thought and commentary on prototyping that follow is credited to observations made by Michael Schrage (2000) in his book *Serious Play.* Anyone interested in Six Sigma should read this book, and specifically the portions dealing with the "innovation and learning" aspects of Six Sigma activities. We do not put quotes around the next two paragraphs, but much of it is paraphrased from Schrage's book.

Whenever an organization wants to enhance its innovative culture and practices, it soon discovers that it must significantly change how it prototypes. One must literally change how people *play* with the prototypes they build. An organization's prototyping methods determine, to a large degree, its ability to profit from innovation. Except for the unforgiving veracity of the marketplace, there is no better way to promote value-creating behavior than through the use of prototypes. And as prototyping media become cheaper and more powerful, it becomes increasingly difficult for management to resist the desire to explore one more design option, to play just a little bit longer. An organization's ability to create customer and shareholder value now depends on its ability to use its prototyping tools effectively. It is not a question of more and better information, or effective management and sharing of the knowledge we already have. Rather, we must determine the kinds of prototypes we need to create in order to create new value for the customer. Some experts are claiming that the number and quality of prototypes an organization develops and interrogates are directly proportional to the ultimate quality of the product under development. And the recent explosion of computational choice and speed, and the array of modeling options at everyone's disposal, make the *organizational culture* the critical success factor in managing innovation.

Virtually all product and service innovation results from the constant struggle between the customer's wish list (specifications) and the prototypes built to articulate those desirable traits. The ongoing corporate dialogue between specifications and prototypes is the heart and soul of innovative design. But even the most technically accurate model cannot guarantee effective communication or productive interaction. There are also profound differences between organizations that build prototypes primarily to *create* questions and those that do so to *answer* questions. Once again, culture has an important impact. As a rule, the technical beauty of a final product is a function of the number of prototypes tested and the prototyping cycles per unit of time. Prototyping cycle time is becoming an important metric in today's competitive environment. This emphasis on time is radically redefining what a good prototyping culture looks like. Some organizations have moved on to "periodic prototyping," which requires design teams to produce a given number of prototypes on a fast, fixed schedule. One suggested indicator of the health and well-being of an internal prototyping culture is to measure the time interval between initial creation of the first prototype and the first "demo" of a subsequent prototype to senior management. *Serious Play* contains more about exploring the intimate relationship between innovation and Six Sigma.

Below is our example of an "innovation" balanced scorecard as defined by Kaplan and Norton (1996). In this case, we are using a commercial firm's new product development process as an example. For the sake of brevity, we offer only one metric per category.

- *Financial:* Percentage of the annual product development capital budget dedicated to acquiring prototyping media and methods.
- *Customer:* Average number of prototypes evaluated by the customer before release of beta-level engineering drawings to production engineering.
- *Internal:* Average number of hours per week that product development engineers spend in the shop building, or supervising the building of, prototypes.
- *Innovation and Learning:* The running-average, annual percentage of the product development staff that make site visits to the prototyping facilities of Fortune 100 companies.

Innovation is, by definition, a risky business. The key to personal innovation is self-confidence (Magaziner and Patinkin, 1989). Napoleon Bonaparte noted that "the art of choosing men is not nearly as difficult as the art of enabling those one has chosen to attain their full worth." Any attempt to reduce the risk of innovation by trying to manage it by committee deals a serious, if not lethal, blow to the advancement of leading-edge technology (Botkin et al., 1984). To mitigate the risk of the innovation process requires frequent and rapid prototyping that involves customer involvement before the glue is even dry on the "artifacts." It is also important to visualize and treat "technology development [innovation] as separate from scientific inquiry" (Carnegie Commission, 1992). Drucker (1985) goes so far as to suggest that systematic measurement of the performance of an organization, as innovator, is mandatory.

14.2 ENCOURAGING CREATIVITY

This section covers only a few observations relative to the encouragement of creativity as a component of a Six Sigma strategy. Creativity can manifest itself in many ways. When we attempt to measure creativity in the context of an organization's ongoing Six Sigma strategy, should we measure creativity, invention, innovation, or something else we haven't thought of yet? Should certain processes or groups be measured on multiple measures of "creativity" simultaneously? Creativity can be demonstrated in a number of domains, including science, writing, drawing, dance, advertising, etc. (Sternberg and Lubart, 1995). Do we have ways of measuring the creativity of potential employees in these multiple domains, and do we have any insight into appropriate job classifications for those demonstrating exceptional creativity in one or more of these domains? Do we want creativity to be an important hiring and performance criterion for the job of janitor? Forklift driver? Secretary? Corporate pilot? Webmaster? Cost accountant? Vice president of Human Resources? Each organization should address these issues as needed.

Next, let us look at the available information on what motivates creativity. We will limit the bulk of our discussion to creativity in the context of organizational research and development. The first reason we are doing this is that, as we've already mentioned, research and development has been a major economic driver in the U.S. economy and others around the globe. Second, we believe that much of the low-hanging fruit that are primary targets for Six Sigma improvement projects can be found in the R&D organization. We regret to say that little information is available on other, nonscience domains in terms of how to motivate creativity. However, we believe that motivators of creativity in the arena of science and product development also apply to other domains. Unfortunately, little work has been done to measure the effect of key R&D motivators and de-motivators on other domains, or vice versa.

The Environment

Considerable research has been conducted on the influence of the environment on creativity. Unfortunately, the results draw no definite conclusions. For every study identifying X as important, one can find another study rejecting X as important. This is most likely due to the modest but important differences in the research methods used to evaluate creativity. Sternberg and Lubart (1995) suggest that this contradictory evidence results from the existence of important, nonlinear variables involved in these studies that interact with one another in such a way that different outcomes are reached, or conclusions made, depending on the magnitude of the variables being studied. For example, supervision (one common factor in these types of studies) can inhibit creativity if its use is excessive, but it can be a motivator if its use is reasonable.

Taylor (1961) interviewed 12 department heads in 10 research laboratories. Ten of the 12 identified the working environment as the most important variable affecting creativity. We hasten to add that department heads may view creativity differently from practicing scientists and engineers. In another study of 165 scientists engaged in R&D (Amabile, 1988), a number of variables were consistently identified as being motivators (9) or inhibitors (11) of creativity; these results are shown in Table 14.2. We have taken the liberty of attempting to align factors in the table that represent opposites of one another. The blank spaces in the table indicate where such "opposites" were not reported. There are many, many lists like this in the literature that contain similar conclusions on this type of research. One aspect of Table 14.2 worth noting is that salary does not show up anywhere on the list. One should not conclude that salary is not important to scientists and engineers. Rather, a firm's technical staff consists of some of the highest-paid non-executive individuals on the payroll. As such, Maslow's theory of the hierarchy of needs would suggest that this specialized group of workers does not view compensation as an unsatisfied need.

TABLE 14.2   Motivators and Inhibitors of R&D Creativity

Motivators	Inhibitors
Freedom and control over work	Lack of freedom
Good management that sets goals, prevents distractions, and is not overly strict	Red tape
Sufficient resources	Insufficient resources
Encouragement of new ideas	Unwillingness to risk change Apathy
Collaboration across divisions	Poor communication Defensiveness within the organization
Recognition of creative work	Poor rewards
Sufficient time to think	Time pressure
Challenging problems	
Sense of urgency for getting work done	
	Competitiveness
	Critical, unrealistic, or inappropriate evaluation

A number of the motivators listed in Table 14.2 are consistent with a Six Sigma strategy. Scientists and engineers want challenging problems, a sense of urgency, collaboration across functions, sufficient resources, encouragement of new ideas, and recognition. All of these motivators should be present on Six Sigma projects. The motivator "sufficient time to think" may pose somewhat of a problem for scientists, but it should be much less of a problem for engineers. Scientists like to read the literature, go to conferences, ponder problems, play in the lab, experiment, and so forth. Those are all appropriate in the pursuit of R&D. However, we are not suggesting that an organization reengineer the R&D process as an S^4 project. Rather, we want scientists to improve the new-product development process. Much of that work will require scientists to work on S^4 projects that focus on the support activities associated with R&D, as delineated earlier.

We will leave this topic with a quotation from Warren Bennis: "There are two ways to be creative. One can sing and dance. Or one can create an environment in which singers and dancers flourish." Six Sigma activity requires significant amounts of both. Creative leaders provide creative environments for all workers, including themselves.

Intelligence

We have been focusing on the effect of the environment on creativity. We will finish this section by briefly commenting on creativity's dependence on the additional factors of intelligence, knowledge, personality, and motivation.

There is considerable evidence that some minimal level of intelligence is necessary for creativity, but intelligence alone is an insufficient condition for creativity (Barron and Harrington, 1981). Standardized tests to measure one's intelligence quotient (IQ) are not reliable predictors of creative ability. They suffer from cultural bias and many other shortcomings. IQ tests may be the best thing we have for screening high school graduates for admission to colleges, but they are not very effective at predicting on-the-job success. The best predictor of future creative behavior in a domain is past creative behavior in that same domain (Barron and Harrington, 1981). An accomplished professional acquaintance of one of the authors is a medical doctor with undergraduate and postdoctoral studies in mathematics. Over decades of professional work in mathematics and modeling, he has observed that the best predictor of one's ability to publish seminal work is that person's past history of such publication (Albanese, 1997).

It is humbling to consider Wiener's (1993) contention that a truly creative idea "is, to a large extent a lucky and unpredictable accident." If this is true, it has tremendous implications for organizational creativity. If senior leaders do not create an environment in which serendipitous events can be pursued by the R&D staff and other creative agents, then many potentially great ideas will be lost—possibly to the competition.

We hasten to interject the concept of emotional intelligence (EQ), as contrasted with the traditional measure of intelligence, IQ. EQ is a relatively new concept to the business community (Goleman, 1997). In the future, a robust economy will provide job opportunities for individuals with IQ as well as those with EQ (Cox and Alm, 1999). Fortunately, both IQ and EQ skills can be enhanced by education and training.

Knowledge

In a knowledge-based society an organization's most important asset is the knowledge worker. There are two types of knowledge, explicit and tacit. Nonaka and Takeuchi (1995), in their book *The Knowledge-Creating Company,* offer insights that go a long way toward creating an environment that is conducive to an S^4 strategy.

In short, the Western world views knowledge as primarily formal and systematic, the stuff of books and journals. By contrast, the Japanese and many other Eastern societies view words and numbers as the tip of the knowledge iceberg. They view knowledge as primarily tacit in nature. And tacit knowledge has two major components: A brain that contains know-how guides the skilled hands of the craftsman. This knowledge is difficult to explain, or document in writing. The second component consists of the internalized mental models and belief systems that shape the way we view the world. To a large extent our worldview colors our vision of what is possible and not possible. One of the challenges facing any serious S^4 implementation is learning how to extract this tacit knowledge from the minds of one's knowledge workers. It is easy to interrogate a process using control charts and designed experiments. It is another thing to extract and share the organizational wisdom that is so essential to the measurement and analysis phases of S^4 implementation.

"What raises living standards over time isn't companies vying for customers in existing industries. It's competition from new goods and services and new production techniques, bringing about new industries, new jobs, and higher living standards" (Cox and Alm, 1999). Only the *accessible* knowledge (and wisdom) of knowledge workers can deliver on this promise.

Personality

It is difficult to decide where to begin on such a nebulous concept as personality, and yet we all know that personality exists, that personality differences exist, and that these differences can be striking. A number of authors define personality as a preferred way of interacting with the environment. The idea that personality is all genetically based and hence unchangeable is flawed. Personality is at least partially under the control of the individual. A person who wants to develop any specific personality trait(s) can be successful if he or she is committed to the task. On the other hand, anyone who has been around an obsessive-compulsive personality knows how "stuck" these individuals seem to be in that mode of behavior. Some people undergo psychotherapy for years in an attempt to deal with what their environment defines as a personality disorder. To the extent that certain mental states are chemically (genetically) based, they may require medical attention. To the extent that they represent personal preferences based on learned behavior, they can be modified with sufficient motivation.

We will begin by presenting the list of "creative" personality traits posited by Sternberg and Lubart (1995); see Table 14.3. We will not discuss this list. We present it so the reader might compare it with Csikszentmihalyi's (1996) list discussed below. One should also consider Margaret Boden's (1991) perspective on creativity: she argues convincingly that any theory of creativity must include a model of conceptual computation by the mind.

TABLE 14.3   Personality Traits of Creative Individuals

PERSONALITY TRAITS (Sternberg and Lubart, 1995)

Perseverance in the face of obstacles
Willingness to take sensible risks
Willingness to grow
A tolerance for ambiguity
Openness to experience
Belief in self and the courage of one's convictions

Our hope is that this overview will help the reader more readily identify truly creative people within organizations. More important, we hope the reader will come to appreciate and value the perspective that these sometimes misunderstood and mislabeled individuals can contribute to an S^4 work environment.

Mihaly Csikszentmihalyi is a university professor who has made important contributions in the area of creativity. He views creativity as one of many properties of a complex system—the mind. Complex systems cannot be explained by observing individual components of the system. Csikszentmihalyi (1996) believes that creative individuals can be identified by the complexity that they exhibit. Less-creative individuals presumably have less of this complexity, possibly none at all. The complexity that creative individuals exhibit shows up in the form of dichotomous behaviors or traits. Whereas most people have personalities that are strongly unipolar, creative people tend to be bipolar in a number of traits. This bipolarity expresses itself in the form of dichotomies. Creative people have personality traits that are complex and confusing to the rest of us because their personalities seem to be dual in nature.

For example, creative people tend to display a great deal of physical energy, and yet at other times are found to be quiet and relaxed. A number of scientists have reported "spontaneously" discovering answers to difficult questions after periods of reflection and contemplation. Some people postulate that these spontaneous insights are a result of parallel brain processing that is occurring during rest as well as work. Most creative people, including nonscientists, have work patterns that allow for periods of physical inactivity and mental contemplation. So one should not make any rash judgments upon entering a laboratory and finding a senior scientist with her feet on the desk, looking out the window with an inquisitive grin. High levels of physical activity at work often impress management. Are these people demonstrating creativity? Or are they producing a lot of undesirable product very quickly? Long-term patterns of performance can provide the answer. Do not be surprised if certain members of S^4 project teams vacillate between frenzied activity and immobility. Some of the problems these S^4 teams undertake demand as much creativity as a Picasso painting.

Creative people tend to be smart and naive at the same time. They seem to be smart in a specific domain but naive in many other areas involving common sense. In the world of Six Sigma, naïveté may be more important than mastery of facts and figures. Frequently a "dumb" question yields the insight needed to solve a problem. S^4 practitioners must be careful in their role as team leaders not to embarrass team members who ask naive questions about statistical matters. The chances are very good that if one team member has a question, the others have similar doubts, concerns, or confusion. Why do we use rational subgroups of size four or five when creating control charts? Why don't factor effects get confused when using full factorial designs involving all possible combinations of factors? Smart people also have the ability to propose novel solutions using knowledge rather than naïveté. This skill of envisioning solutions or representations of problems is found in many individuals who have the ability to think both divergently and convergently. Convergent thinking is the traditional mode taught in school, focusing on finding the one right (best?) answer. Divergent thinking looks for multiple solutions to a single problem before winnowing the field to just one answer. There is considerable evidence suggesting that S^4 project teams will come up with better (more creative) solutions if they are asked to provide multiple solutions (divergent thinking) rather than a single optimum solution (convergent thinking).

Creative people possess the wonderful duality of playfulness and discipline. The playfulness is a reflection of the childlike innocence these people possess. Creative people withhold judgment, because in their innocence there is no sin. It is this playful, nonjudgmental behavior that gives creative people the capacity to see possibilities that have been lobotomized in others. Do not be surprised if the fun-loving jokester on an S^4 project team is also the one who quietly works alone until midnight trying to figure out what the data really mean. Give team members the freedom to be occasionally sarcastic or emphatically goal oriented. Sarcasm, after all, can be an ironic form of humor. And effective goal setting is demanded by most S^4 activity.

Creative individuals can be highly imaginative or deeply grounded in reality. A distinguishing characteristic between creative and noncreative people is that creative people are good at coming up with highly original ideas that are not bizarre. Normal people's ideas are rarely original but often bizarre. And as with all of the dichotomies we are discussing, creative people have to be able to switch between modes with relative ease.

Creative people seem to be both extroverted and introverted at the same time. If one desires to stand apart from the pack in any field of endeavor, one has to study, emulate, and converse with the current giants in that field. Consider how you will work with both introverts and extroverts on an S^4 project team. Also consider how to effectively harness the creativity of the rare introverted extrovert on the team. As we mentioned before, creativity is scarce and most people tend to be at one of the two extremes on this dichotomous scale. The creative personality tends to be a rare flower, a hybrid. Allen (1977) reports some fascinating findings on the probability of communication between scientists as a function of office separation distance. He reports that the likelihood of weekly contact between two scientists within the same work group rapidly approaches zero as their separation distance exceeds a mere 30 feet. More alarming is the finding that the sharing of laboratory space does not enhance intergroup communication. It seems that laboratories are introversion zones, whereas offices are extroversion zones. The message here for S^4 project leaders is to co-locate the team for the duration of the project if at all possible.

Creative individuals are both humble and proud at the same time. Their appreciation of the contributions of others is sincere. On the other hand, creative people recognize that their talents are indeed rare, and that compared to the mass of humanity they have contributed significantly, even if in a very narrow domain. S^4 practitioners need to be extremely careful in handling the double-edged sword of pride and humility. Our advice is to display humility by controlling pride; prideful displays are appropriate only in the context of team accomplishment.

Creative people have internalized the best attributes of both males and females. Csikszentmihalyi (1996) reports that creative and talented girls are more dominant and tough than other girls, and that creative boys are more sensitive and less aggressive than their male peers are. He labels this phenomenon "psychological androgyny" and rejects it as a tendency toward homosexuality. The value of these additional cross-gender attributes is that they empower members of both sexes to interact with their environment equipped with a larger arsenal of responses. Again, not every man or woman on an S^4 project team merits the label "creative." What are the implications when teams are fortunate enough to have one or more creative contributors as members? The major advantage is less contention and more cohesion among all team members.

Motivation

Motivation is defined as the driving force that leads someone to action. Pelz and Andrews (1976) interviewed scientists in an attempt to identify their primary sources of motivation for engaging in the pursuit of science and discovery. One interesting finding was that high performers were inclined toward a scientific discipline, not lifelong employment as an "organization person." Those scientists who relied on their own ideas for motivation were highly effective. Those who relied on their supervisors for stimulation were below-average performers. And lastly, these scientists reported much higher motivation when engaged in "broad mapping of new areas" as opposed to probing "deeply into a narrow area."

Koestler (1964) claims that man not only responds to the environment but also interacts with it by asking questions. He claims man's primary motivations for prodding and probing the environment are novelty, surprise, conflict, and uncertainty.

Shekerjian (1990) makes the interesting observation that the greater one's mastery of a discipline, the less motivation there is to generate new approaches to problem solving. This suggests that regular, if not frequent, transitions into new disciplines should be highly motivational. It is certainly consistent with Koestler's point that man loves to tinker in areas rich in novelty and surprise. Could it be that world-class "familiarity" breeds confidence, expertise, status, and laziness?

Gretz and Drozdeck (1992) assert that most people are motivated by needs, and that the three needs that most commonly drive human behavior are the needs for power, association with others, and achievement.

So what should motivate S^4 project teams to work diligently to increase customer satisfaction and improve organizational profitability? First we must make the point that an S^4 strategy is, above all else, a major organizational investment in people. The infrastructure requirements to enable and maintain an effective S^4 strategy are driven by investments in human capital. Cox and Alm (1999) claim that "the rewards of investing in human capital go to the workers themselves in the form of fatter paychecks and to companies in the form of higher productivity and greater profits." As global competition increases and the rate of technological change accelerates, there will be tremendous social turmoil as the uneducated and unprepared secede from the expanding economy. Those with the capability and desire to keep pace with radical change will find virtually limitless opportunity for variety, growth, and prosperity.

LIST OF SYMBOLS

Symbols used locally in the book are not shown.

ABC	Activity-based costing
AFR	Average failure rate
ANOM	Analysis of means
ANOVA	Analysis of variance
AQL	Accept quality level
ARL	Average run length
ASQ	American Society for Quality (previously ASQC, American Society for Quality Control)
c (chart)	Control chart for nonconformities
CEO	Chief executive officer
CFO	Chief financial officer
CL	Center line in a control chart
COPQ	Cost of poor quality
CTC	Critical to cost
CTD	Critical to delivery
CTP	Critical to process
CTQ	Critical to quality (similar to KPOV)
CTV	Critical to value

C_p	Capability index [AIAG 1995] (In practice, some calculate using "short-term" standard deviation, while others calculate using "long-term" standard deviation.)
C_{pk}	Capability index [AIAG 1995] (In practice, some calculate using "short-term" standard deviation, while others calculate using "long-term" standard deviation.)
C&E	Cause-and-effect (diagram)
DOE	Design of experiments
DMAIC	Define-Measure-Analyze-Improve-Control
DPMO	Defects per million opportunities
ECO	Engineering change order
FMEA	Failure mode and effects analysis
Gage R&R	Gage repeatability and reproducibility
ID	Interrelationship digraph
IM	Information management
JIT	Just in time
KCA	Knowledge-centered activity
KPIV	Key process input variable
KPOV	Key process output variable
LCL	Lower control limit of control chart
LSL	Lower specification limit
MBO	Management by objectives
MSA	Measurement systems analysis
ML	Maximum likelihood (estimate)
MTBF	Mean time between failures
NPP	Normal probability plot
NSF	National Science Foundation
NTF	No trouble found
np (chart)	Control chart of number of nonconforming items
ORT	Ongoing reliability test
p (chart)	Control chart of fraction nonconforming
P_p	Performance index [AIAG 1995] (calculated using "long-term" standard deviation)
P_{pk}	Performance index [AIAG 1995] (calculated using "long-term" standard deviation)
QFD	Quality function deployment
ROI	Return on investment
RTY	Rolled throughput yield
R	Range in control chart

R&D	Research and development
s	Standard deviation of a sample
SBU	Strategic business unit
SPC	Statistical process control
S^4	Smarter Six Sigma Solutions
TQM	Total quality management
UCL	Upper control limit of a control chart
URL	Uniform Resource Locator (www address)
USL	Upper specification limit
u (chart)	SPC chart of number of nonconformities per unit
VP	Vice president of an organization
www	World Wide Web
XmR (chart)	Control chart of individual and moving range measurements
$\bar{x}$	Mean of a variable x
$\bar{x}$ chart	Control chart of means (i.e., x-bar chart)
$\tilde{x}$	Median of variable x
μ	Mu, population true mean
σ	Sigma, population standard deviation

GLOSSARY

From Breyfogle (1999), *Implementing Six Sigma,* plus additional terms.

"30,000-foot level" control chart: For a continuous response, an *XmR* chart of one random sample that is taken infrequently—for example, daily. The time between samples should be long enough to span normal short-term noise to the process. For an attribute response, a control chart should be used that captures between-subgroups variability within the control limits. The time to collect attribute data needs to be long enough to span normal short-term noise to the process.

Accuracy: The extent to which the observed value or calculated average value of a measurement agrees with the "true" value.

Alpha (α) **risk**: Risk of rejecting the null hypothesis erroneously. It is also called type I error or producer's risk.

Alternative hypothesis (H_a): *See* Hypothesis testing.

Analysis of means (ANOM): An approach to compare means of groups of common size n to the grand mean.

Analysis of variance (ANOVA): A statistical procedure that can be used to determine the significant effects in a factorial experiment.

Attribute data: Numerical information that is collected using a "nominal" scale of measurement. The term *nominal* indicates the data are associated with "names" or "categories." A typical example of attribute data is political party: Republican, Democrat, Independent. Other examples are good vs. bad, complete vs. not complete, works vs. does not work. A typical quality example of attribute data is Pass or Fail. Often referred to as discrete data.

Baseline: The current output response of a process.

Beta (β) risk: Risk of not rejecting the null hypothesis erroneously. Also called type II error or consumer's risk.

Bias: The difference between the observed average of measurements and the reference value. Bias is often referred to as accuracy.

Bimodal distribution: A distribution that is a combination of two different distributions resulting in two distinct peaks.

Binomial distribution: A distribution that is useful to describe discrete variables or attributes that have two possible outcomes—for example, a pass/fail proportion test, heads/tails outcome from flipping a coin, or defect/no defect present.

Box plot: A technique that is useful to pictorially describe various aspects of data—for example, between-part and within-part.

Brainstorming: Consensus building among experts about a problem or issue using group discussion.

Cause-and-effect diagram (C&E diagram): This is a technique that is useful in problem solving using brainstorming sessions. With this technique possible causes from such sources as materials, equipment, methods, and personnel are typically identified as a starting point to begin discussion. The technique is sometimes called an Ishikawa diagram or fishbone diagram.

Cause-and-effect matrix: A tool that can aid with the prioritization of key input variables.

_c_ chart: _See_ control charts.

Chronic problem: A description of the situation where a process control chart may be in control; however, the overall response is not satisfactory. Common cause variability within a process yields an unsatisfactory response. For example, a manufacturing process has a consistent "yield" over time; however, the average yield number is not satisfactory.

Common cause variation: A source of variation that is random; an inherent natural source of variation.

Confidence interval: The region containing the confidence limits or band of a parameter where the bounds are large enough to contain the true parameter value. The bands can be single-sided, describing an upper/lower limit, or double-sided, describing both upper and lower limits.

Continuous distribution: A distribution used in describing the probability of a response when the output is continuous. _See_ Response.

Continuous response: _See_ Response.

Control: The term "in control" is used in process control charting to describe when the process indicates that there are no special causes. "Out of control" indicates that there is/are special cause(s).

Control chart: A statistical tool used to track an important condition over time and watch for changes in both the average and the variation of that condition. There are two general types of control charting to use, depending on the type of data being plotted. The two types of data are attribute (count) data and variables (continuous) data. Attribute data are typically associated with a p chart, np chart, c chart, and u chart. Variables data are associated with XmR charts, $\bar{x}$ and R charts, and $\bar{x}$ and S charts. See Chapter 10 of Breyfogle (1999), *Implementing Six Sigma,* for more details on how to select the proper control chart for a given situation and type of data.

Control plan: A description of the inputs to a process that should be monitored or error-proofed for the purpose of maintaining satisfactory output for a KPOV.

Correlation: The determination of the effect of one variable upon another in a dependent situation.

Correlation coefficient (r): A statistic that describes the strength of a linear relationship between two variables is the sample correlation coefficient. A correlation coefficient can take values between -1 and $+1$. A -1 indicates perfect negative correlation, while a $+1$ indicates perfect positive correlation. A zero indicates no correlation.

Cost of poor quality: Cost-of-quality issues often are given the broad categories of internal failure costs, external failure costs, appraisal costs, and prevention costs.

Critical to quality (CTQ): A term widely used by General Electric in its Six Sigma activities, which describes an element of a design, characteristic of a part, or attribute of a service that is critical to quality in the eyes of the customer. Formerly known as key quality characteristics. Similar to the KPOV term used within this book. Within GE, sometimes referred to as "Y."

Customer: Someone for whom work or a service is performed. The end user of a product is a customer of the employees within a company that manufactures the product. There are also internal customers in a company. When an employee does work or performs a service for someone else in the company, the person who receives this work is a customer of this employee.

Deductive reasoning: The act of reasoning that starts from the general and moves toward the more specific. Working from general knowledge, it is often possible to hypothesize more detailed or specific actions, outcomes, effects, or consequences that would flow from this general knowledge. The act of engaging in this kind of thinking is called deductive reasoning. *See also* Inductive reasoning.

Defect: A failure to meet an imposed requirement on a single quality characteristic, or a single instance of nonconformance to the specification.

Defective: A unit of product containing one or more defects.

Defects per million (DPM): A defect rate expressed as the number of defect occurrences observed per million units produced. Also referred to as the parts-per-million (ppm) defect rate.

Defects per million opportunities (DPMO): Rather than calculating defect rates based on the number of final units that pass or fail testing, some organizations estimate how many opportunities there are for failure in a given final product. When this approach is used, DPMO is defined as the number of defects counted, divided by the actual number of opportunities to make a defect, then multiplied by one million. This measure can be converted directly into a sigma quality level.

Defects per unit (DPU): The number of defects counted, divided by the number of products or characteristics produced.

Degrees of freedom (df or v): Number of measurements that are independently available for estimating a population parameter. For a random sample from a population, the number of degrees of freedom is equal to the sample size minus one.

Design of experiments (DOE): A method of experimentation in which purposeful changes are made to the inputs to a process (factors) in order to observe the corresponding changes in the outputs (responses).

Design for manufacturing: A set of design techniques for creating products that can be produced and assembled easily, thus improving quality while reducing cost.

Design for Six Sigma (DFSS): Designing and creating a component, system, or process with the intent of meeting and/or exceeding all the needs of customers and KPOV requirements upon initial release. The goal of DFSS is that there be no manufacturing issues with the design upon the initial release of the design to manufacturing.

Distribution: A pattern that randomly collected numbers from a population follow. The normal, Weibull, Poisson, binomial, and log-normal distributions discussed in this book are applicable to the modeling of various industrial situations.

DOE: *See* Design of experiments.

Effect: That which is produced by a cause.

Entitlement: The expected performance level of a product or process when its major sources of variation are identified and controlled.

Error (experimental): Ambiguities during data analysis from such sources as measurement bias, random measurement error, and mistake.

Evolutionary operations (EVOP): An analytical approach whereby process conditions are changed structurally in a manufacturing process (e.g., using a fractional factorial experiment design matrix) for the purpose of analytically determining changes to make for product improvement.

Experiment: A process undertaken to determine something that is not already known.

Experimental error: Variations in the experimental response under identical test conditions. Also called "residual error."

Factors: Variables that are varied to different levels within a factorial designed experiment or response surface experiment.

F test: A statistical test that utilizes tabular values from the F distribution to assess significance.

Failure mode and effects analysis (FMEA): A process in which each potential failure mode in every sub-item of an item is analyzed to determine its effect on other sub-items, and on the required function of the item itself. Opposite of fault tree analysis.

Failure rate: Failures/unit time or failures/units of usage (i.e., 1/MTBF). Sample failure rates are: 0.002 failures/hour, 0.0003 failures/auto miles traveled, 0.01 failures/1,000 parts manufactured. Failure rate criterion (ρ_a) is a failure rate value that is not to be exceeded in a product. Tests to determine if a failure rate criterion is met can be fixed or sequential in duration. With fixed-length test plans, the test design failure rate (ρ_t) is the sample failure rate that cannot be exceeded in order to certify the criterion (ρ_a) at the boundary of the desired confidence level. With sequential test plans, failure rates ρ_1 and ρ_0 are used to determine the test plans.

Firefighting: An expression that describes the performing of emergency fixes to problems that recur.

Fishbone diagram: *See* Cause-and-effect diagram.

FMEA: *See* Failure mode and effects analysis.

Fractional factorial experiment: A designed experiment strategy that assesses several factors/variables simultaneously in one test, where only a partial set of all possible combinations of factor levels are tested to more efficiently identify important factors. This type of test is much more efficient than a traditional one-at-a-time test strategy.

Full factorial experiment: Factorial experiment where all combinations of factor levels are tested.

Gage: Any device used to obtain measurements. The term is frequently used to refer specifically to shop floor devices, including go/no-go devices.

Gage Repeatability & Reproducibility (Gage R&R): The evaluation of a measurement system to determine the variation inherent in the measurement equipment and the appraiser (measurer). As a general rule of thumb, Gage R&R variability should be a factor of 10 smaller than the variability being measured in a part or process.

Gage Repeatability & Reproducibility (R&R) study: The evaluation of measuring instruments to determine capability to yield a precise response. Gage repeatability is the variation in measurements considering one part and one operator. Gage reproducibility is the variation between operators measuring one part.

Gantt chart: A visual representation of a time sequence of activities. Gantt charts consist of a number of staggered straight lines, each of which represents the start and stop time of a given activity. It is essentially a planning tool for managing projects. This graphic form is attributed to Henry L. Gantt, a World War I military adviser.

Hidden factory: Reworks within an organization that have no value and are often not considered within the metrics of a factory.

Histogram: A graphical representation of the sample frequency distribution that describes the occurrence of grouped items.

Hypothesis testing: Consists of a null hypothesis (H_0) and alternative hypothesis (H_a) where, for example, a null hypothesis indicates equality between two process outputs and an alternative hypothesis indicates nonequality. Through a hypothesis test a decision is made whether to reject a null hypothesis or not reject a null hypothesis. When a null hypothesis is rejected, there is a risk of error. Most typically, there is no risk assignment when we fail to reject the null hypothesis. However, an appropriate sample size could be determined such that failure to reject the null hypothesis is made with β risk of error.

In control: The description of a process where variation is consistent over time and only common causes exist.

Inductive reasoning: The act of reasoning that starts from the specific and moves toward the general. Working with a number of specific facts or details, it is often possible to hypothesize a more general or macroscopic view of how some aspect of the world works. The act of engaging in this kind of thinking is called inductive reasoning. *See also* Deductive reasoning.

Interaction: Two factors are said to interact if one factor's effect on a response variable is dependent on the level of the other factor. This concept can be generalized to third-order and higher interactions. However, even three-factor interactions are relatively rare in nature.

Key process input variable (KPIV): Inputs that are determined to be important to the overall output of a process. GE often referred to this as an "X."

Key process output variable (KPOV): Process outputs that are important to meeting the needs of customers. Within GE, these variables are considered as critical-to-quality (CTQ) variables, sometimes using a "Y" nomenclature.

Lower control limit (LCL): For control charts the LCL is the lower limit below which a process statistic (mean, range, standard deviation, etc.) is considered to be "out of control" due to some assignable cause. Dr. Shewhart (*see* Shewhart Control Charts) observed empirically that three standard deviations below the process mean or median is the most appropriate value for the LCL.

Lower specification limit (LSL): The lowest value of a product dimension or measurement that is acceptable to the customer.

Main effect: An estimate of the effect of a factor measured independently of other factors.

Management by fact: Management that bases decisions and actions on verifiable data rather than by instinct, opinion, or hunch.

Mean: The mean of a sample ($\bar{x}$) is the sum of all the responses divided by the sample size. The mean of a population (μ) is the sum of all responses of the population divided by the population size. In a random sample of a population, $\bar{x}$ is an estimate of μ for the population.

Measurement systems: The complete process of obtaining measurements. This includes the collection of equipment, operations, procedures, software, and personnel that affects the assignment of a number to a measurement characteristic.

Measurement systems analysis: *See* Gage repeatability and reproducibility (R&R).

Minitab: Statistical analysis software that is often used within Six Sigma training.

Multi-vari chart: A chart that is constructed to display the variance within units, between units, between samples, and between lots.

Nonconformity: A condition observed within a product or process that does not conform to one or more customer specifications. Often referred to as a defect. A nonconforming part or process may contain more than one nonconformity.

Normal distribution: A bell-shaped distribution that is often useful to describe various physical, mechanical, electrical, and chemical properties.

Normal variation: *See* Common cause variation.

np **chart:** *See* Control chart.

Null hypothesis (H_0): *See* Hypothesis testing.

One-at-a-time experiment: An individual tries to fix a problem by making a change and then executing a test. Depending on the findings, something else may need to be tried. This cycle is repeated indefinitely.

Outlier: A data point that does not fit a model because of an erroneous reading or some other abnormal situation.

Pareto chart: A graphical technique used to quantify problems so that effort can be expended in fixing the "vital few" causes, as opposed to the "trivial many." Named after Wufredo Pareto, a European economist.

Pareto principle: Eighty percent (the trivial many) of the trouble results from 20% (the vital few) of the problems.

Passive data analysis: Data are collected and analyzed as the process is currently performing. Process alterations are not assessed.

p **chart:** *See* Control chart.

Point estimate: An estimate calculated from sample data without a confidence interval.

Population: The totality of items under consideration. Often referred to as the "universe."

Probability: The chance of something's happening.

Problem solving: The process of determining the cause from a symptom and then choosing an action to improve a process or product.

Problem statement: A statement that describes in specific and measurable terms what is wrong, and the impact potential if the problem is not fixed.

Process: A method to make or do something that involves a number of steps.

Process capability: A normal range of performance of a process when it is operating in a state of statistical process control.

Process capability indices (C_p and C_{pk}): C_p is a measurement of the allowable tolerance spread divided by the actual 6σ data spread. C_{pk} has a ratio similar to that of C_p except that this ratio considers the shift of the mean relative to the central specification target.

Process flow diagram (chart): Path of steps of work used to produce or do something.

Process spread: The range of values that a given product or process characteristic displays. This term most often applies to the range, but it could also be used in relation to variance.

Qualitative factor: A factor that has discrete levels. For example, product origination with factor levels of supplier A, supplier B, and supplier C.

Quantitative factor: A factor that is continuous. For example, a product can be manufactured with a process temperature factor between 50°C and 80°C.

Quality function deployment (QFD): A technique that is used, for example, to get the "voice of the customer" in the design of a product.

Random effects (or components of variance) model: An experiment where the variance of factors is investigated (as opposed to a fixed effects model).

Random causes: *See* Common cause variation.

Range: For a set of numbers, the absolute difference between the largest and smallest value.

Regression analysis: A statistical technique for determining the relationship between one response and one or more independent variables.

Repeatability: The variation in measurements obtained with one measurement instrument when used several times by one appraiser while measuring the identical characteristic on the same part.

Replication: Test trials that are made under identical conditions.

Representative sample: A limited group of observations drawn from a much larger population or universe in such a way that the observable characteristics of the sample (mean, mode, standard deviation, skewness, range, etc.) reflect the character and diversity of the original population being sampled.

Reproducibility: The variation in the average of the measurements made by different appraisers using the same measuring instrument when measuring identical characteristics on the same part.

Response: In this book, three basic types of responses (i.e., outputs) are addressed: continuous, attribute, and logic pass/fail. A response is said to be continuous if any value can be taken between limits (e.g., 2, 2.0001, and 3.00005). A response is said to be attribute if the evaluation takes on a pass/fail proportion output (e.g., 999 out of 1,000 sheets of paper on the average can be fed through a copier without a jam). In this book, a response is said to be logic pass/fail if combinational considerations are involved that are said to always cause an event to either pass or fail (e.g., a computer display design will not work in combination with a particular keyboard design and software package).

Response surface methodology (RSM): The empirical study of relationships between one or more responses and input variable factors. The technique is used to determine the "best" set of input variables to optimize a response and/or gain a better understanding of the overall system response.

Robust: A description of a procedure that is not sensitive to deviations from some of its underlying assumptions.

Rolled throughput yield (RTY): The yield of individual process steps multiplied together is the RTY of the overall process. RTY reflects the "hidden factory" rework issues of a process.

Run chart: The plot of a variable over time.

Sample size: The number of observations made or the number of items taken from a population.

Sampling distribution: A distribution derived from a parent distribution by random sampling.

Shewhart Control Charts: Dr. Walter A. Shewhart, of Bell Telephone Laboratories, developed control charts that tracked common cause variability and identified when special causes occurred within processes.

Sigma: The Greek letter (σ) that is often used to describe the standard deviation of data. It is a measure of the consistency of a process.

Sigma level or sigma quality level: A quality that is calculated by some to describe the capability of a process to meet specification. A six sigma quality level is said to have a 3.4 ppm defect rate. Pat Spagon from Motorola University prefers to distinguish between sigma as a measure of spread and sigma as a measure of product or process quality level (Spagon, 1998).

Significance: A statistical statement indicating that the level of a factor causes a difference in a response with a certain degree of risk of being in error.

Single-factor analysis of variance: One-way analysis of variance with two levels (or treatments) that is to determine if there is a significant difference between level effects.

Six Sigma: A term coined by Motorola that emphasizes the improvement of processes for the purpose of reducing variability and making general improvements.

Smarter Six Sigma Solutions (S^4): Term used within this book to describe the wise and often unique application of statistical techniques to create meaningful measurements and effective improvements.

Smarter Six Sigma Solutions Assessment (S^4 Assessment): Using statistically based concepts while determining the "best" question to answer from the point of view of the customer. Assessment is made to determine if the right measurements and the right actions are being conducted. This includes noting that there are usually better questions to ask to protect the customer than "What sample do I need?" or "What one thing should I do next to fix this problem?" (i.e., a one-at-a-time approach). S^4 resolution may involve putting together what often traditionally are considered "separate statistical techniques" in a "smart" fashion to address various problems.

Soft savings: Savings that are a result of cost avoidance or efficiency improvements where there are no immediate tangible results. Examples include the reduction of development cycle time and improved operation efficiency when there is no immediate head-count reduction since a person is still currently working at the same operation.

Soft skills: A person who effectively facilitates meetings and works well with other people has good "soft skills."

Special causes: *See* Sporadic problem.

Specification: A criterion that is to be met by a part or product.

Specification limits: *See* Upper specification limit; Lower specification limit.

Sporadic problem: A problem that occurs in a process because of an unusual condition (i.e., from special causes). An out-of-control condition in a process control chart.

Stable process: A process that is in statistical control.

Standard deviation (σ, s): A mathematical quantity that describes the variability of a response. It equals the square root of variance. The standard deviation of a sample (s) is used to estimate the standard deviation of a population (σ).

Statistical control: A condition that describes a process that is free of special causes of variation. Such a condition is most often portrayed on a control chart.

Statistical process control (SPC): The application of statistical techniques in the control of processes. SPC is often considered a subset of SQC, where the emphasis in SPC is on the tools associated with the process but not product acceptance techniques.

Trend chart: A chart to view the resultant effect of a known variable on the response of a process.

Trial: One of the factor combinations in an experiment.

Trivial many: *See* Pareto principle.

t **test:** A statistical test that utilizes tabular values from the Student's "t" distribution to assess, for example, whether two population means are different.

Type I error: *See* Alpha risk.

Type II error: *See* Beta risk.

Type III error: Answering the wrong question.

u **chart:** *See* Control chart.

UCL: *See* Upper control limit.

Uncertainty (Δ): An acceptable or desirable amount of change from a criterion. The parameter is used when considering β risk in sample size calculations.

Upper control limit (UCL): For control charts, the UCL is the upper limit above which a process statistic (mean, range, standard deviation, etc.) is considered to be "out of control" due to some special cause. Dr. Shewhart (*see* Shewhart Control Charts) observed empirically that three standard deviations above the process mean (or median) is the most appropriate value for the UCL.

Upper specification limit (USL): The highest value of a product dimension or measurement that is acceptable to the customer.

Variables data: Data that can assume a range of numerical responses on a continuous scale, as opposed to data that can assume only discrete levels. Often referred to as continuous data.

Variance (σ^2, s^2): A measure of dispersion of observations based upon the mean of the squared deviations from the arithmetic mean.

Variance components analysis: A statistical technique to determine and quantify the sources of variance to a process.

Vital few: *See* Pareto principle.

Wave: A Six Sigma term referring to a four-week training course that is typically conducted over a four-month time period.

Weibull distribution: This distribution has a density function that has many possible shapes. The shape parameter (b) and the location parameter (k) describe the two-parameter distribution. This distribution has an x-intercept value at the low end of the distribution that approaches zero, which yields a zero probability of a lower value. The three-parameter has, in addition to the other parameters, the location parameter (x_0), which is the lowest x-intercept value.

XmR **charts:** Also known as Individuals Moving Range (*ImR* charts). *See* Control chart.

$\bar{x}$ **& R charts:** *See* Control chart.

$\bar{x}$ **& S charts:** *See* Control chart.

"X's": The designation for those variables that are the root causes of effects observed in products or processes ("explanatory variables" in regression analysis). S^4 focuses on measuring and improving X's so as to promote subsequent improvements in Y's. Also called independent variables, or factor(s) in a designed experiment.

"Y's": The designation for those variables that are the observed effects or outcomes resulting from changes to the input variables affecting a product or process. We generally assume that these Y's are the result of one or more X's expressing themselves in a product or process. Also called dependent variables, or response(s) in a designed experiment.

Yield, Final: The ratio of acceptable product at the end of a process to the initial number product that was started initially with the process. Often this ratio is multiplied by 100 and expressed as a percentage value. For example, a yield of 0.857 equates to an 85.7% yield.

Yield, Rolled Throughput: *See* Rolled throughput yield.

REFERENCES

AIAG (1995), *Statistical Process Control (SPC) Reference Manual,* Chrysler Corporation, Ford Motor Company, General Motors Corporation.

Albanese, Richard A. (1997), personal communication.

Allen, Thomas J. (1977), *Managing the Flow of Technology,* MIT Press, Cambridge, MA.

Amabile, T.M. (1988), A Model of Creativity and Innovation in Organizations, *Research in Organizational Behavior* 10: 123–167.

American Productivity & Quality Center (1988), *Measuring White Collar Work,* Houston, TX.

Aviation (1998), "Success with Six Sigma Often an Elusive Goal," *Aviation Week and Space Technology*, vol. 139, no. 20, p. 53.

Barron, F. and Harrington, D.M. (1981), Creativity, Intelligence, and Personality. *Annual Review of Psychology* 32: 439–476.

Bergquist, Timothy M. and Ramsing, Kenneth D. (September 1999), Measuring Performance After Meeting Award Criteria: Study Compares Perceived Success to Financial Data of Award Winners and Applicants. *Quality Progress* 32(9): 66–72.

Bisgaard, Soren (1991), Teaching Statistics to Engineers. *The American Statistician* 45(4): pp. 274–283.

Blakeslee, Jerome A. Jr. (July 1999), Implementing the Six Sigma Solution: How to Achieve Quantum Leaps in Quality and Competitiveness. Quality Progress 32(7): 77–85.

Boden, Margaret A. (1991), *The Creative Mind,* Basic Books, New York, NY.

Botkin, James, Dimancescu, Dan, and Stata, Ray (1984), *The Innovators,* Harper & Row, New York, NY.

Box, George (1991), *Teaching Engineers Experimental Design With a Paper Helicopter*, CQPI Report 76, Center for Quality and Productivity Improvement, University of Wisconsin at Madison, Madison, WI.

Box, G.E.P., Bisgaard, S. and Fung, C. (1988), An Explanation and Critique of Taguchi's Contributions to Quality Engineering, *Quality and Reliability Engineering International*, 4(2): 123–131.

Box, G.E.P., Hunter, W.G., and Hunter, S.J. (1978), *Statistics for Experimenters*, John Wiley, New York, NY.

Bouckaert, Geert, and Balk, Walter (Winter 1991), Public Productivity Measurement: Disease and Cures, *Public Productivity & Management Review*, pp. 229–235.

Boyett, Joseph H., Kearney, A. T., and Conn, Henry P. (1992), What's Wrong with Total Quality Management, *Tapping the Network Journal,* pp. 10–14.

Bradstreet, Thomas E. (1996), Teaching Introductory Statistics Courses So That Nonstatisticians Experience Statistical Reasoning, *The American Statistician* 50(1): 69–78.

Breyfogle, Forrest W. III (1999), *Implementing Six Sigma: Smarter Solutions Using Statistical Methods,* John Wiley & Sons, New York, NY.

British Deming Association (1991), *A System of Profound Knowledge (Number 9),* SPC Press, Knoxville, TN.

British Deming Association (1990), *Profound Knowledge (Number 6),* SPC Press, Knoxville, TN.

Brooks, Frederick P. Jr. (1975), *The Mythical Mon-Month: Essays on Software Engineering,* Addison-Wesley, Reading, MA.

Brown, Mark G., and Svenson, Raynold A. (July/August 1988), Measuring R&D Productivity, *Research Technology Management*, pp. 39–43.

Bryce, G. Rex (1992), Data Driven Experiences in an Introductory Statistics Course for Engineers Using Student Collected Data, *1992 Proceedings of the Section on Statistical Education*, pp. 155–160, American Statistical Association, Alexandria, VA.

Buderi, Robert (Jan. 18, 1993), American Inventors are Reinventing Themselves, *Business Week,* pp. 78–82.

Carnegie Commission on Science, Technology, and Government (December 1992), *Environmental Research and Development: Strengthening the Federal Infrastructure,* p. 42.

Chang, Richard Y. (1994), *Creating High-Impact Training,* Richard Chang Associates, Irvine, CA.

Chappell, Tom (1999), *Managing Upside Down, The Seven Intentions of Values-Centered Leadership,* William Morrow and Company, Inc., New York.

Cheek, Tom (1992), *Six Sigma Black Belt Certification*, Signed by A. W. Wiggenhorn, President, Motorola University, and Mikel Harry, Ph.D., Corporate Director Six Sigma Research Institute, September 25, 1992.

Cobb, George (1991), Teaching Statistics: More Data, Less Lecturing, *Amstat News*, vol. 182, pp. 1, 4.

Collier, D.W. and Gee, R.E. (May 1973), A Simple Approach to Post-Evaluation of Research, *Research Management,* pp. 12–17.

Conway, William E. (1999), personal communications, Conway Management Company, Nashua, NH.

Conway, William E. (1994), *Winning the War on Waste: Changing the Way We Work,* Conway Quality, Nashua, NH.

Conway, William E. (1992), *The Quality Secret: The Right Way to Manage,* Conway Quality, Nashua, NH.

Cox, W. Michael and Alm, Richard (1999), *Myths of Rich & Poor,* Basic Books, New York, NY.

Crosby, Philip (1995), *Quality Without Tears*, McGraw-Hill, New York, NY.

Crosby, Philip B. (1999), personal communications: www.philipcrosby.com/philspage/philsbio.htm.

Crosby, Philip B. (1992), *Completeness: Quality for the 21st Century*, Plume Books, New York, NY.

Crosby, Philip B. (1984), *Quality Without Tears: The Art of Hassle-Free Management*, Plume Books, New York, NY.

Crosby, Philip B. (1979), *Quality Is Free: the Art of Making Quality Certain*, Mentor Books, New York, NY.

Csikszentmihalyi, Mihaly (1996), *Creativity: Flow and the Psychology of Discovery and Invention,* HarperCollins, 1996.

Cupello, James M. (September–October 1999), Training Technologists in Experimental Design, *Research Technology Management* 42(5): 47–55.

de Bono, Edward (1992), *Serious Creativity: Using the Power of Lateral Thinking to Create New Ideas,* HarperCollins, New York, NY.

DeLavigne, Kenneth T. and Robertson, J. Daniel (1994), *Deming's Profound Changes,* Prentice Hall, Englewood Cliffs, NJ.

Deming, W. Edwards (1986), *Out of the Crisis,* MIT Center for Advanced Engineering Study, Cambridge, MA, pp. 23–24, 97–98.

Denecke, Johan (1998), 6 Sigma and Lean Synergy, *Allied Signal Black Belt Symposium*, AlliedSignal Inc., pp. 1–16.

Dertouzos, Michael L. (1990), *Made In America,* HarperPerennial, New York, NY.

Draman, Rexnord H. and Chakravorty, Satya S. (2000), An Evaluation of Quality Improvement Project Selection Alternatives, *Quality Management Journal* 7(1): 58–73.

Drucker, Peter F. (1985), *Innovation and Entrepreneurship,* Harper & Row, New York, NY.

Edwards, Shirley A. and McCarrey, Michael W. (January 1973), Measuring the Performance of Researchers, *Research Management*, pp. 34–41.

Fagan, Matt (2000), personal 3M communications.

Felix, Glenn H. (1983), *Productivity Measurement With the Objectives Matrix,* Oregon Productivity Center Press, Oregon State University, Corvallis, OR.

Floberg, Brandi (December 1999), 1999 Statistical Process Control Software Buyers Guide, *Quality Digest* 19(12): 51–64.

Folkman, Joe (1998), *Employee Surveys That Make a Difference, Using Customized Feedback Tools to Transform Your Organization*, Executive Excellence Publishing, Provo, UT.

Franklin, LeRoy A., Cooley, Belva J., and Elrod, Gary (October 1999), Comparing the Importance of Variation and Mean of a Distribution, *Quality Progress* 32(10): 90–94.

Gabel, Stanley H. and Callanan, Terry (May 19, 1999), Six Sigma at Eastman Kodak Company, 1999 ASA Quality and Productivity Conference, American Statistical Association, Schenectady, NY.

General Electric (1997), *General Electric Company 1997 Annual Report.*

General Electric (1996), *General Electric Company 1996 Annual Report.*

Gilman, John J. (July 1993), personal communication.

Gilman, John J. (1992), *Inventivity: The Art and Science of Research Management,* Van Nostrand Reinhold, New York, NY.

Gitlow, Howard S. and Gitlow, Shelly J. (1987), *The Deming Guide to Quality and Competitive Position,* Prentice-Hall, Englewood Cliffs, NJ.

Godfrey, A. Blanton (May 19, 1999), Six Sigma Quality: From Strategic Deployment to Bottom-Line Results, 1999 ASA Quality and Productivity Conference, American Statistical Association, Schenectady, NY.

Godfrey, A. Blanton (December 1999), Building a Scorecard, *Quality Digest* 19(12): 16.

Goleman, Daniel P. (1997), *Emotional Intelligence: Why It Can Matter More Than IQ,* Bantam Books, New York, NY.

Gomory, Ralph (June 1990), Essays: Of Ladders, Cycles and Economic Growth, *Scientific American,* p. 140.

Gretz, Karl F. and Drozdeck, Steven R. (1992), *Empowering Innovative People,* Probus Publishing, Chicago, IL.

Gunter, Bert (1993), Through a Funnel Slowly with Ball Bearing and Insight to Teach Experiment Design, *The American Statistician* 47(4): 265–269.

Hahn, Gerald J., Hill, William J., Hoerl, Roger W., and Zinkgraf, Stephen A. (August 1999), The Impact of Six Sigma—A Glimpse Into the Future of Statistics, *American Statistician* 53(3): 208–215.

Harry, Mikel J. (February 2000a), The Quality Twilight Zone, *Quality Progress* 33(2): 68–71.

Harry, Mikel J. (January 2000b), A New Definition Aims to Connect Quality Performance with Financial Performance, *Quality Progress* 33(1): 64–66.

Harry, Mikel and Schroeder, Richard (2000), *Six Sigma: The Breakthrough Management Strategy Revolutionizing the World's Top Corporations,* Currency, New York, NY.

Hendrix, C.D. (April 1991), Signal-to-Noise Ratios: A Wasted Effort, *Quality Progress,* pp. 75–76.

Hiebeler, Robert, Kelly, Thomas B., Ketteman, Charles (1998), *Best Practices, Building Your Business with Customer Focused Solutions*, Arthur & Anderson, Simon & Schuster, New York.

Hogg, Robert V. (1991), Statistical Education: Improvements Are Badly Needed, *The American Statistician* 45(4): 342–343.

Hughes, Richard L., Ginnett, Robert C., and Curphy, Gordon J. (1993), *Leadership: Enhancing the Lessons of Experience,* Richard D. Irwin, Homewood, IL.

Hughes, Richard L., Ginnett, Robert C., and Curphy, Gordon J. (1993), *Leadership— Enhancing the Lessons of Experience,* Richard D. Irwin, Boston, MA.

Jonash, Ronald S. and Sommerlatte, Tom (1999), *The Innovation Premium,* Arthur D. Little, Inc., Perseus Books, Reading, MA.

Jones, D. (1998), Firms Air for Six Sigma Efficiency, *USA Today,* July 21, 1998 Money Section.

Juran, Joseph M. (1999a), personal communications: www.juran.com/drjuran/bio_jmj.html.

Juran, Joseph M. and Godfrey, A. Blanton (1999b), *Juran's Quality Handbook* (Fifth Edition), McGraw-Hill, New York, NY.

Juran, J.M. and Gryna, Frank M. (1988), *Juran's Quality Control Handbook* (Fourth Edition), McGraw-Hill, New York, NY.

Juran, J.M. (1992), *Juran on Quality by Design: The New Steps for Planning Quality into Goods and Services,* Free Press, New York, NY.

Juran, J.M. (1989), *Juran on Leadership for Quality: An Executive Handbook,* Free Press, New York, NY.

Juran, J.M. (1988), *Juran on Planning for Quality,* Free Press, New York, NY.

Juran, J.M. (1964), *Managerial Breakthrough,* McGraw-Hill, New York, NY.

Jusko, Jill (6 Dec 1999), A Look at Lean, *Industry Week.*

Kaplan, Robert S. and Norton, David P. (1996), *The Balanced Scorecard,* Harvard Business School Press, Boston, MA.

Kaplan, Robert S. and Norton, David P. (Sep/Oct 1993), Putting the Balanced Scorecard to Work, *Harvard Business Review* 71(5): 138–140.

Kaplan, Robert S. and Norton, David P. (Jan/Feb 1992), The Balanced Scorecard— Measures That Drive Performance, *Harvard Business Review* 70(1): 71–79.

Kiemele, M.J. (1998), Information presented in this paragraph was contributed by M.J. Kiemele, Ph.D., of Air Academy Associates.

King, B. (1987), *Better Designs in Half the Time, Implementing QFD in America,* Goal/QPC, Methuen, MA.

Koestler, Arthur (1964), *The Act of Creation,* Viking Penguin, New York, NY.

Kuhn, Thomas S. (1970), *The Structure of Scientific Revolutions* (Second Edition, Enlarged), University of Chicago Press, Chicago, IL.

Latzko, William J. and Saunders, David M. (1995), *Four Days with Dr. Deming,* Addison-Wesley, Reading, MA.

Levitt, Theodore (1991), *Thinking About Management,* Free Press, New York, NY.

Lobbestael, Wayne and Vasquez, Bud (May–June 1991), Measure to Improve: Ideas on Implementing Measurement, *Program Manager,* pp. 39–44.

Lowe, J. (1998), *Jack Welch Speaks,* John Wiley & Sons, New York, NY.

Lubar, S. (1990), New, Useful, and Nonobvious, *Invention & Technology* 6: 8.

Lundin, Barbara L. Vergetis (October 1999), Presenting the Software Showcase, *Quality Progress* 32(10): 95–117.

Magaziner, Ira and Patinkin (1989), *The Silent War,* Vintage Books, New York, NY.

Messina, W.S. (1987), *Statistical Quality Control for Manufacturing Managers,* Wiley, New York.

Michalko, Michael (1998), *Cracking Creativity: The Secrets of Creative Genius,* Ten Speed Press, Berkeley, CA.

Micklethwait, John and Wooldridge, Adrian (1997), *The Witch Doctors: Making Sense of the Management Gurus,* Random House, New York, NY.

Miles, Maguire (1999), *Cowboy Quality,* Quality Progress, vol. 32, no. 10, Oct 1999, pp. 27–34.

Minitab, Inc. (2000), 3081 Enterprise Dr., State College, PA 16801, http://www.minitab.com.

Montgomery, D.C. (1990–1991), Using Fractional Factorial Designs for Robust Process Development, *Quality Engineering* 3(2): 193–205.

Moore, David S. (1992), Teaching Statistics as a Respectable Subject, in *Statistics for the Twenty-First Century* (Florence and Sheldon Gordon, editors)*,* Mathematical Association of America (MAA) Notes, vol. 26, pp. 14–25, Mathematical Association of America.

National Science Foundation (1989), *Science and Engineering Indicators 1998,* National Science Board, Arlington, VA.

Neave, Henry R. (1990), *The Deming Dimension,* SPC Press, Knoxville, TN.

Nonaka, Ikujiro and Takeuchi, Hirotaka (1995), *The Knowledge-Creating Company: How Japanese Companies Create the Dynamics of Innovation,* Oxford University Press, New York, NY.

O'Dell, Carla and Grayson, C. Jackson Jr. (1998), *If Only We Knew What We Know,* Free Press, New York, NY.

Paton, Scott M. (February 1993), Four Days with W. Edwards Deming, *Quality Digest,* pp. 23–27.

Pelz, Donald C. and Andrews, Frank M. (1976), *Scientists in Organizations,* Institute for Social Research, University of Michigan, Ann Arbor, MI.

Perez-Wilson, Mario (1999), *Six Sigma: Understanding the Concepts, Implications and Challenges,* Advanced System Consultants, Scottsdale, AZ.

Price, Frank (1990), *Right Every Time: Using the Deming Approach,* Marcel Dekker, New York, NY.

Pritchett, Price and Pound, Ron (1997), *The Employee Handbook for Organizational Change,* Pritchett Publishing, Dallas, TX.

Pyzdek, Thomas (1999), *The Complete Guide to Six Sigma,* p. 431, Quality Publishing, Tucson, AZ.

Pyzdek, Thomas (January 2000), Six Sigma and Lean Production, *Quality Digest,* p. 14.

Pyzdek, Thomas (1999a), *The Complete Guide to Six Sigma,* Quality Publishing, Tucson, AZ.

Pyzdek, Thomas (November 1999b), Why Six Sigma Is Not Enough, *Quality Digest,* p. 26.

Pyzdek, Thomas (June 1999c), Six Sigma Is Primarily a Management Program, *Quality Digest,* p. 26.

Quality Sourcebook (January 2000), *Quality Digest* 20(1): 109–127.

Robbins, Anthony (1991), *Awaken the Giant Within,* Fireside, NY.

Ross, P.J. (1988), *Taguchi Techniques for Quality Engineering,* McGraw-Hill, New York, NY.

Roussel, Philip A. (1991), *Third Generation R&D,* Harvard Business School Press, Boston, MA.

Scherkenbach, William W. (1988), *The Deming Route to Quality and Productivity: Road Maps and Roadblocks,* ASQC Quality Press, Milwaukee, WI.

Schiemann, William A. and Lingle, John H. (1999), *BULLSEYE!: Hitting Your Strategic Targets Through High-Impact Measurement,* Free Press, New York, NY.

Schmidt, Stephen R. (1993), *Personal Communication,* Design of Experiments Workshop, University of Texas at Austin, Austin, TX.

Schmidt, S.R., Kiemele, M.J. and Berdine, R.J. (1997), *Knowledge Based Management,* Air Academy Press & Associates, Colorado Springs, CO.

Schmidt, Warren H. and Finnegan, Jerome P. (1992), *The Race Without a Finish Line,* Jossey-Bass, San Francisco, CA, p. 309.

Schrage, Michael (2000), *Serious Play,* Harvard Business School Press, Boston, MA.

Schrage, Michael (1990), *Shared Minds,* Random House, New York, NY.

Senge, P. M. (1990), *The Fifth Discipline: The Art and Practice of the Learning Organization,* Doubleday/Current, New York, NY.

Shekerjian, Denise (1990), *Uncommon Genius: How Great Ideas Are Born,* Viking, New York, NY.

Slater, Robert (1999), *The GE Way Fieldbook,* McGraw-Hill Companies, Inc., New York, NY.

Slater, Robert (2000), *The GE Way Fieldbook: Jack Welch's Battle Plan for Corporate Revolution,* McGraw-Hill, New York, NY.

Snee, Ronald D. (September 1999a), Why Should Statisticians Pay Attention to Six Sigma?: An Examination for Their Role in the Six Sigma Methodology, *Quality Progress* 32(9): 100–103.

Snee, Ronald D. (May 19, 1999b), Why Should We Pay Attention to Six Sigma, 1999 ASA Quality and Productivity Conference, American Statistical Association, Schenectady, NY.

Stahl, Michael J. and Steger, Joseph A. (January 1977), Measuring Innovation and Productivity: A Peer Rating Approach, *Research Management,* pp. 35–38.

Sternberg, Robert J. and Lubart, Todd I. (1995), *Defying the Crowd: Cultivating Creativity In a Culture of Conformity,* Free Press, New York, NY.

Stookey, S.D. (January 1980), The Pioneering Researcher and the Corporation, *Research Management,* pp. 15–18.

Taguchi, Genichi (1999), personal communications: www.amsup.com/TAGUCHI.

Taguchi, G. and Konishi, S. (1987), *Taguchi Methods, Orthogonal Arrays and Linear Graphics*, American Supplier Institute, Dearborn, MI.

Tatsuno, Sheridan M. (1990), *Created in Japan,* Harper Business, New York, NY.

Taylor, D.W. (October 1961), Environment and Creativity, in Conference on the Creative Person, Institute of Personality Assessment and Research, University of California at Berkeley, Berkeley, CA.

Texas Instruments (1992), *What Is Six Sigma,* TI-29077, F21110.

Thor, Carl (August 1989), Taking the Measure of White-Collar Productivity, *Manufacturing Engineering,* pp. 44–48.

Thor, Carl (1988), A Complete Productivity and Quality Measurement Management System, *Manager's Notebook,* American Productivity and Quality Center, Houston, TX.

Tomasko, Robert M. (1990), *Downsizing,* American Management Association (AMA), AMACOM, New York, NY.

Vaill, Peter B., original source unknown.

Wadsworth, Harrison M. (1990), *Handbook of Statistical Methods for Engineers and Scientists*, McGraw-Hill, New York, NY, pp. 19.21, 19.24.

Walton, Mary (1986), The Parable of the Red Beads, *The Deming Management Method,* Perigee Books, New York, NY, pp. 40–51.

Watts, Donald G. (1991), Why Is Introductory Statistics Difficult to Learn? And What Can We Do to Make It Easier? *The American Statistician* 45(4): 290–291.

Weaver, Richard G. and Farrell, John D. (1999), *Managers As Facilitators,* Berrett Koehler Publishers, Inc., San Francisco, CA.

Web Consulting Honolulu (1999), http://www.sitetuneup.net.

Wheeler, Donald J. (1995), Advanced Topics in Statistical Process Control, SPC Press, Knoxville, TN.

Wiener, Norbert (1993), *Invention: The Care and Feeding of Ideas,* MIT Press, Cambridge, MA.

Wise, R.I. (1999), A Methodology for Aligning Process Level and Strategy Level Performance Metrics, ASQ Quality Management Forum, Spring 1999.

Wortman, Bill (1995), *The Quality Engineer Primer* (Fourth Edition), Quality Council of Indiana, West Terre Haute, IN.

Wortman, William L. (December 15, 1999), personal communications, Quality Council of Indiana, West Terre Haute, IN. Original source: anonymous.

www.46sigma.com/faq.htm (December 21, 1999).

Zinkgraf, Stephen A. and Snee, Ronald D. (Sigma Breakthrough Technologies, Inc.) (May 20, 1999), Institutionalizing Six Sigma in Large Corporations: A Leadership Roadmap, 1999 ASA Quality and Productivity Conference, American Statistical Association, Schenectady, NY.

INDEX

Accounts payable, overdue, 90–92
Action items on change, 133
Akers, John, 36
Albanese, Richard A., 233
Allen, Thomas J., 237
AlliedSignal, 21, 27, 35, 36
Alm, Richard, 233, 238
Amabile, T.M., 231
American Productivity and Quality Center, 94, 95
American Society for Quality, 15
Analysis of means (ANOM), 64, 90, 91–92
Analysis of variance (ANOVA), 19, 20, 64, 65, 91–92, 139, 187, 200, 217
Andrews, Frank M., 224, 228, 231
ANOM. *See* Analysis of means
ANOVA. *See* Analysis of variance
Applied research (invention), 226–227
Appraisal costs, 49
AT&T Statistical Quality Control Handbook, 12
Attribute data, 65, 79, 91, 196
Aviation Week and Space Technology, 136, 137

Baker, James, 137
Balanced scorecard approach, 97–98, 155, 185, 197, 214, 222
Balanced Scorecard, The (Kaplan and Norton), 97
Balk, Walter, 94
Barron, F., 233
Bennis, Warren, 232
Berquist, Timothy M., 15
Black Belts (project manager/facilitator), x, 27, 33, 34, 35, 111, 113, 124, 126, 127, 135
    number of, 136
    roles and responsibilities, 125, 165
    selection of, 128–129
    training of, 145
Blakeslee, Jerome A., 25, 150
Bodek, Norman, 14–15
Boden, Margaret, 234
Bossidy, Larry, 35, 36
Botkin, James, 230
Bottom-line benefits, 105, 109
Bouckaert, Geert, 94
Box, G.E.P., 90

Box-Cox Transformation, 90–91
Box plots, 19, 64, 186, 199, 217
Boyett, Joseph H., 15
Brainstorming, 60–61, 117, 121, 134, 159–160, 188, 192, 196, 214
Breyfogle, Forrest W. III. *See Implementing Six Sigma*
Brown, Mark G., 96
Buderi, Robert, 227
Burn-in, 184

Callanan, Terry, 33
Cause-and-effect diagram, 18, 61, 65, 159–160, 162, 172, 175
    development applications, 216, 220
    manufacturing applications, 186, 188
    service/transactional applications, 198
Cause-and-effect matrix, 17, 19, 61, 151, 160–161, 172, 176
    development applications, 216, 217, 219–220
    manufacturing applications, 196
    service/transactional applications, 197, 198, 206–207
Chakravorty, Satya S., 142
Champions (executive-level manager), 27, 28, 111, 127, 128, 135
    roles and responsibilities, 124
    training, 146
Change orders, 85–87
Chappell, Tom, 132
Charter, project, 161, 163–166
Cheek, Tom, x, 27
Collier, D.W., 224
Common cause variation, 58–59, 60, 61, 79, 82, 159, 188, 191, 195
Communications
    Four Cs of effective, 134–135
    intranet sites in, 179
    report-outs to management, 111–112, 174–179
    among team members, 170
Computer development, 219–220
Computer software, 28, 140, 146–147
Continuous flow production, 22

Continuous response data, 61–65, 79, 91, 183, 188
Control charts, 18, 90, 143
    attribute data, 65, 66, 79, 91
    classic approach to, 62
    continuous data, 61–65, 79, 91, 183
    development applications, 215, 219
    at 50-foot level, 60, 65, 200
    manufacturing applications, 183–184, 185, 187, 189, 191–193
    and project selection, 156
    rule number 1, 82
    service/transactional applications, 195–196, 198, 200–201, 204, 205, 206
    Shewhart model, 193
    short run, 192–193
    at 30,000-foot level, 20–21, 59–65, 66, 79, 80, 156, 175, 185, 198, 215, 219
    three-way, 191–192
Control measures, 94, 95
Control plan, 17, 20, 187, 200, 218
Conway, William E., 11, 47
COPQ. *See* Cost of poor quality
Correlation, 187, 200, 217
Cost of doing nothing, 45, 46, 50, 102
Cost of poor quality (COPQ), 18, 45, 47–50, 91, 98, 155, 157, 185, 191, 197, 214
Cox, W. Michael, 233, 238
Creativity, 230–231
    balanced scorecard, 226
    components of, 224
    and environment, 231–232
    and innovation, 224–225, 226, 228
    and intelligence, 233
    and invention, 226–227
    and knowledge, 233–234
    and motivation, 237–238
    and personality, 234–237
    in Six Sigma, 131, 225–226
Critical-to-quality (CTQ) characteristics, 95, 143, 151
Crosby, Philip B., 12–13, 171
Cross-functional metrics, 153
Csikszentmihalyi, Mihaly, 234, 235, 237
CTQ. *See* Critical-to-quality characteristics

Culture, 130–134, 221–223
    customer focus of, 132–133
    and employee motivation, 132
    empowered, 132
    force field analysis of, 130–131
    information sources on, 223
    and invention, 226–227
    stimulating change in, 133–134
    See also Creativity; Innovation
Cupello, James M., 47
Current control plans, 19
Customer-centered metrics, 153
Customers
    complaints, 133
    failures, 191
    feedback loops, 119, 121
    hidden needs of, 222
    satisfaction, 6, 45, 95, 98, 121–123, 132–
        133
    setup problems, 219
    specifications, 20, 83, 229
    surveys, 118–119, 120
Cycle time, 4, 74, 193, 215–216, 229

Data collection and analysis, 55–69, 101
    and goal setting, 65, 67–68
    and out-of-specification conditions, 55–59
    passive, 172
    process control charting, 59–65
    sampling plans, 156
    See also Metrics; Statistical tools
Data entry coordinator, 166
Defectives, 78–79, 188
Defect rate, 41, 46, 62, 74, 79, 155
Defects, 41, 183, 191, 196
Defects per million opportunities (DPMO),
        71, 75–76, 102, 143, 190
Defects-per-opportunity (DPO), 75, 190
Define-Measure-Analyze-Improve-Control
        (DMAIC), 104, 171
Delivery time, 22, 193, 200–204
Deming, W. Edwards, 8–11, 12, 58
Deming Prize, 9
Demonstration runs, 187, 200
Denecke, Johan, 21, 22

Dertouzos, Michael L., 225
Designed for Six Sigma (DFSS). *See*
        Development applications
Design of experiments (DOE), 17, 19, 20, 65,
        68, 85, 145, 148, 172
    development applications, 211, 212, 213,
        214, 217, 218, 219–220
    manufacturing applications, 183–184, 187,
        188, 193–194
    service/transactional applications, 195,
        196, 197, 200, 207–208
Design for manufacturability (DFM), 98
Development
    cycle time, 74–75, 215–216, 225
    innovation phase, 228, 230
    See also Development applications
Development applications, 211–220
    computer, 219–220
    examples of, 211–212
    process measurements, 212–213
    21-step integration of tools, 213–219
DFSS. *See* Development applications
Disney World, 133
Dodge, Harold, 11
DOE. *See* Design of experiments
Dot plots, 157, 177
DPMO. *See* Defects per million opportunities
DPO. *See* Defects-per-opportunity
Draman, Rexnord H., 142
Drozdeck, Steven R., 238
Drucker, Peter F., 12, 224, 225, 228, 230

Eastman Kodak, 33–35
ECO. *See* Engineering change orders
Edwards, Shirley A., 228
Emotional intelligence (EQ), 233
Employees
    communication with, 134–135
    full-time, 29
    opinion survey data, 87–90
    and performance measurement, 94, 96
    skills, 29–30
    traits, 29
    See also Black Belts; Motivation; Teams;
        Training

Empowerment, 132
Engineering change order (ECO), 85–87
Engineering process control (EPC), 193
EPC. *See* Engineering process control
EWMA. *See* Exponentially weighted moving average
Executives
 communications, 111–112, 134–135
 leadership, 115, 117, 135
 project report-outs to, 111–112, 174–179
 roles and responsibilities, 124, 126
 training, 146
Expense reduction, 191
Expertise, 224
Exponentially weighted moving average (EWMA), 193

Facilitators, 128–129, 160, 167
Failure costs, 49
Failure modes and effects analysis (FMEA), 17, 19, 61, 65, 172, 176
 development applications, 213, 216–217, 219, 220
 manufacturing applications, 186
 service/transactional applications, 199
Failure rate, 184, 190
Family of measures concept, 95, 96, 98
Farrell, John, 128, 130, 133, 167
Feedback, 170–171
50-foot level control chart, 60, 65, 200
Financial measures, 95
Financial records, of quality costs, 47–48
Finnigan, Jerome P., 46
Firefighting activities, 1, 58, 59, 60, 79, 102, 185, 188, 198
"Five whys," 112
Floberg, Brandi, 147
Flowcharts, process, 158–159, 185, 186, 198, 216
FMEA. *See* Failure modes and effects analysis
Folkman, Joe, 118, 119
Force field analysis, 130–131
Fortune, 221
Fourteen Points, Deming's, 9–10
Frequency of occurrence units, 215

Functional test effectiveness, 191
Functional test yields, 191

Gabel, Stanley H., 33
Gage repeatability & reproducibility (Gage R&R), 85, 139
Galvin, Bob, 32
Gantt charts, 166, 168–169, 185, 198, 214
Gaussian curve, 38–39
Gee, R.E., 224
General Electric (GE)
 applications of Six Sigma, 6
 benefits of Six Sigma, 32–33, 35
 communications, 135
 customer surveys, 118
 deployment strategy, 112–113
 employee motivation, 132
 executive leadership, 117, 135
 implementation lessons, 136
 market capitalization, 32
 performance measurement, 95, 97–98
 process capability, 74
 project execution, 171
 project selection, 151
 project-tracking database, 129–130
 roles and responsibilities, 27, 124
 success factors, 115–116
 training, 145
Germaine, Jack, 32
Gilman, John J., 224, 225
Gitlow, Howard S. and Shelly J., 9
Goals
 arbitrary, 65, 67, 170
 of executive meetings, 117
 meaningful, 170
 as parent projects, 158
 setting, 67–68
 SMART, 17, 65, 136
 strategic, 121–122, 136
 training, 140
Godfrey, A. Blanton, 36, 98
Goleman, Daniel P., 233
Gomory, Ralph, 224
Government specifications on product compliance, 81–85
Grayson, C. Jackson, Jr., 55

Green Belts, x, 27, 33, 113, 126, 146
Gretz, Karl F., 238

Hahn, Gerald J., 145
Harrington, D.M., 233
Harry, Mikel J., x, 15, 16, 27, 124
Harvard Business Review, 97
Hidden factory, 24–25, 76–77, 96, 190
Hidden needs, customer's, 222
Hiebeler, Robert, 133
Histogram, 91, 157
Hollans, Steve, 136
House of Quality process, 121–123, 150,
    153
Hughes, Richard L., 224
Hypothesis tests, 19, 186, 199, 217

IBM, 36
Imaginitive thinking, 224
Implementation
    common errors, 137
    correctness of, ix–x, 36–37, 46
    elements of plan, 109–113
    models for, 27
    of prioritized projects, 102–6
    through projects *vs.* individual tools, 106–
        107
    selection of Six Sigma provider, 107–109
    traditional approach to, 106
    See also Infrastructure; Training
Implementing Six Sigma (Breyfogle), 6, 17,
        31, 39, 56, 61, 64, 71, 73, 82, 85, 92,
        122, 143, 145, 151, 155, 156, 159, 179,
        184, 191, 193, 194, 196, 197, 201, 207,
        208, 214, 220
Infrastructure, 115–137
    communications in, 111–112, 134–135
    culture of, 130–134
    customer focus of, 118–121
    executive leadership in, 115, 117, 135
    and metrics, 129–130, 136–137
    organization chart, 126, 127
    roles and responsibilities, 111, 124–129,
        136
    strategic goals of, 121–122, 136
    success factors for, 115–116

Innovation
    balanced scorecard, 230
    categories of, 228
    and creativity, 224–225, 226, 228
    culture of, 221
    defined, 224, 227–228
    and productivity, 228–229
    and prototyping, 229, 230
    and technology, 228
Instructors, 28–29
Intelligence, and creativity, 233
Intelligence quotient (IQ), 233
Interrelationship diagram (ID), 115–116
Intranet site, 179
Invention (applied research), 224, 226–227
Inventory accuracy, 191

Japan, quality programs in
    and Deming, 8–9
    and Juran, 12
    and Shingo, 14–15
    and Taguchi, 13
Japanese Union of Science and Engineering
        (JUSE), 8, 12
Jonash, Ronald S., 221
Jones, D., 35
Juran, Joseph M., 11–12, 224
Juran Institute, 36
Jusco, Jill, 21

Kaizen, 224
Kano, Noritaki, 221
Kaplan, Robert S., 97, 98, 226, 227, 230
Key process input variable (KPIV), 19, 95,
        160, 173
    development applications, 217, 218
    manufacturing applications, 186, 187
    service/transactional applications, 199,
        200, 206–207
Key process output variable (KPOV), 18, 19,
        50, 61, 95, 102, 105, 143, 160, 173,
        215
    development applications, 214, 215, 217,
        218, 219
    manufacturing applications, 185, 186,
        187

Key process output variable (KPOV)
  (continued)
  service/transactional applications, 197,
    198, 199
Kiemele, M.J., 36
King, B., 221
Knowledge-centered activity (KCA), x
Knowledge-Creating Company, The (Nonaka
    and Takeuchi), 233
Knowledge worker, 233–234
Koestler, Arthur, 238
Konishi, S., 13
Kotter, John, 137
Kozol, Jonathan, 142
KPIV. *See* Key process input variable
KPOV. *See* Key process output variable
Kuhn, Thomas, 225

Lean manufacturing, 21–25
Levitt, Theodore, 59
Likert scale, 87, 89, 119
Lingle, John H., 28
Lockheed Martin, 35–36
Lombardi, Vince, 68
Lowe, J., x, 27, 33
Lower specification limit (LSL), 72
"Low-hanging fruit," 161, 171–172, 197, 219
Lubart, Todd I., 226, 230, 231, 234, 235
Lundin, Barbara L., 147

McCarrey, Michael W., 228
Magaziner, Ira, 227, 230
Malcolm Baldrige National Quality Award
    (MBNQA), 15, 32
Management. *See* Executives
Managers as Facilitators (Weaver and
    Farrell), 128
Managing Upside Down (Chappell), 132
Manufacturing, lean, 21–25
Manufacturing applications, 6, 183–209
  between- and within-part variability, 191–
    192
  engineering process control (EPC), 193
  examples of, 183–184
  new equipment settings, 193–194

  process improvement strategy,
    188–191
  short runs, 192–193
  21-step integration of tools, 184–188
Map, process, 17, 18, 60, 158–159, 175, 185,
    198, 216
Marginal plots, 64
Market capitalization, 32
Maslow, Abraham, 232
Master Black Belts (statistical expert), 27,
    111, 127, 128
  roles and responsibilities, 124–125
  training, 146
Meadows, Becki, 115, 179
Measurement systems analysis (Gage R&R),
    17, 19, 20, 186, 196, 199, 217
Messina, W.S., 189
Metrics, 3–5, 71–92, 101
  accounts payable data, 90–92
  balanced scorecard, 97–98, 155, 185, 197,
    214, 222
  change order data, 85–87
  counterproductive results, 71
  cross-functional, 153
  customer-centered, 153
  DPMO rate, 75–76, 102, 143, 190
  DPO rate, 75, 190
  employee opinion survey data, 87–90
  in implementation of business strategy,
    129–130, 136–137
  informational, 153, 155
  process capability, 71–74
  process cycle time, 74
  product compliance data, 81–85
  product *vs.* process, 190
  quality levels, 41, 42–43
  rolled throughput yield (RTY), 76–78, 79,
    143, 185, 190, 191, 197, 214
  selection of, 78–79, 152–155
  and service/transactional applications,
    196–197
  training, 143–144
Micklethwait, John, 8
Microsoft Office, 28, 140
Miles, Maguire, 117
Minitab, 28, 140, 147
Monetary measures, 96–97, 143–144

Motivation
    of creativity, 224, 237–238
    and reward, 132, 231–232
    team, 166–171, 238
Motorola, 27, 31–32
Motorola University, 31, 32
Multiple regression, 139
Multi-vari charts, 17, 19, 20, 64, 65, 186, 199, 217
Myths of Six Sigma, 6–8

Napoleon Bonaparte, 230
National Science Foundation (NSF), 226, 227–228
Neave, Henry R., 9
Needs, motivating, 231–232, 238
Needs assessment, 45–51
    "doing nothing" costs, 50
    quality costs, 47–50
    questions for, 50–51
Net profit, 95
New equipment settings, optimizing, 193–194
Noise, lean manufacturing approach to, 22–23
Nonaka, Ikujiro, 233
Noncompliance, 189
Nonconformance, 159, 185, 189, 198, 199, 216
Normal distribution, 38–39
Normal probability plots, 157, 195
Norton, David P., 97, 98, 226, 227, 230

O'Dell, Carla, 55
Opinion surveys, employee, 87–90
Organization chart, 126, 127
Organization culture. *See* Culture
Out-of-specification conditions, 55–59

Pareto charts, 18, 19, 65, 75, 79, 83, 85, 87, 90, 153
    development applications, 216, 219
    manufacturing applications, 184, 186, 188, 189
    service/transactional applications, 195, 196, 198, 201, 203, 204

Pareto principle, 12
Pass/fail functional testing, 217
Pass/fail requirements, 196
Patinkin, 227, 230
p chart, 188, 201, 202, 204
Pelz, Donald C., 224, 228, 237
Performance measurement, 93–98
    balanced scorecard approach to, 97–98
    family of measures approach to, 95, 96, 98
    principles of, 94–97
    types of, 93–94
Personality, creative, 234–237
Personnel. *See* Employees; Training
Planning, 79, 93, 161
Poka-yoke devices, 14, 85
Postmortem discussions, 112
Pound, Ron, 142
PowerPoint, 140
ppm rate, 42–43, 190–191
Prevention costs, 49
Price, Frank, 228
Price, and waste elimination, 47
Price to earnings (P/E) ratio, 95
Pritchett, Price, 142
Probability plot, 83, 84, 87, 89, 90
Process capability, 17, 20, 62, 63, 71–74, 73, 175
Process capability/performance strategies, 139, 143, 157
    development applications, 219
    manufacturing applications, 184, 185
    service/transactional applications, 195, 196, 198, 200
Process control charts. *See* Control charts
Process cycle time, 4, 74, 193, 215–216, 229
Process flowcharts, 158–159, 185, 186, 198, 216
Process mapping, 17, 18, 60, 158–159, 175, 185, 198, 216
Process owner (manager of process), 27, 126, 127, 128, 165
Process yield, 76–78
Product compliance, 81–85
Product development. *See* Development
Productivity, 10, 224, 228–229
Project charter, 161, 163–166
Project execution, 171–173

Project manager, 165
Project plan, 79, 161
Project reports, 111–112, 135, 143, 174–179, 187, 200, 219
Project scoping, 158–161, 171
Project selection, 142–143, 149–150
    assessment, 150–151
    baseline/bottom-line benefits, 155–157
    from "child projects," 158
    and metrics, 152–155
Prototyping, 224, 229
Purchase orders, 193, 195, 202
Pyzdek, Thomas, x, 8, 15, 21, 27, 33

QFD. *See* Quality function deployment
Quality Control Handbook, 12
Quality costs, 47–50
Quality Digest, 147
Quality engineering, Taguchi system of, 13
Quality function deployment (QFD), 121–123, 160, 214, 220
Quality Is Free (Crosby), 13
Quality leader, 128
Quality levels, sigma, 16, 25, 39–40
    comparison among, 26
    and defect rate, 41
    and DPMO rate, 75–76
    implications of, 46–47
    manufacturing application, 190
    metrics, 41, 42–43, 155
    shift value, 39–40
Quality Progress, 147

Ramsing, Kenneth D., 15
Random sampling, 195–196
Range chart, 191
Raytheon, 35
Regression analysis, 20, 65, 187, 188, 200, 217
Reliability tests, 218
Reports, project, 111–112, 135, 143, 174–179, 187, 200, 219
Research and development (R&D), 225, 226, 227, 231, 232
    See also Development; Innovation

Response
    attribute, 65, 79, 91, 196
    continuous, 61–65, 79, 91, 183, 188, 196
Response surface methods (RSM) analysis, 19, 20, 139, 187, 217
Restraining forces on change, 133
Return on investment (ROI), 8, 26–27, 95, 132, 155, 208, 215
"Right Way to Manage, The," 11
Robbins, Anthony, 134
Robust design techniques, 13
Rolled throughput yield (RTY), 18, 76–78, 79, 143, 185, 190, 191, 197, 214
Ross, P.J., 13
Roussel, Philip A., 225
RTY. *See* Rolled throughput yield
Run charts, 18, 185, 188, 198, 215
Rushing, 171

Sampling
    frequency, 60, 185, 198
    plans, 156
    random, 195–196
Scherkenbach, William W., 9
Schiemann, William A., 28
Schmidt, Warren H., 46
Schrage, Michael, 228, 229
Screening measures, 93–94
Senge, Peter, 3
Serious Play (Schrage), 229
Service/transactional applications, 195–209
    delivery time, 200–204
    examples of, 195–196, 208–209
    and metrics, 196–197
    21-step integration of tools, 197–200
    website traffic, 205–208
Service/transactional process, 6
    and process capability metrics, 73–74
    sigma quality level within, 76
Setup time, 23
Seven Deadly Diseases, Deming's, 10
Shareholder rate of return, 221
Shekerjian, Denise, 238
Shewhart, Walter, 11
Shewhart model for control charting, 193
Shingo, Shigeo, 14–15

Short-run charts, 192–193
Sigma
    defined, 25
    See also Quality levels, sigma
Six Sigma
    applications. *See* Development applica-
        tions; Manufacturing applications;
        Service/transactional applications
    benefits of, 16–17, 26–27, 33, 35–36
    characteristics of, 16
    comparison with TQM, 8–16
    and creativity, 131, 225–226
    defined, 5–6, 25, 150
    deployment of. *See* Implementation;
        Infrastructure; Teams; Training
    at Eastman Kodak, 33–35
    at General Electric. *See* General Electric
        (GE)
    and lean manufacturing, 21–25
    at Motorola, 31–32
    myths of, 6–8
    and quality levels, 25, 26
    roles and responsibilities, 27–28, 124–129
    statistical definition of, 37–43
    tools. *See* Metrics; Statistical tools;
        specific tool
Slater, Robert, x, 27, 97, 132, 135, 136
SMART goals (simple, measurable, agreed to,
        reasonable, and time- based), 17, 65,
        136
Snee, Ronald, 16, 17
Software, 28, 140, 146–147
Solution tree, 179, 180
Sommerlatte, Tom, 221
SPC. *See* Statistical process control
Special cause variation, 58–59, 60, 62, 82, 83
Specification limits, 37, 38, 56, 72, 83, 85,
        157, 196
Specification values, 76
Sponsor/process owner (manager of process),
        27, 126, 127, 128
Stahl, Michael J., 228
Standard deviation, 25, 72, 73
Statistical definition of Six Sigma, 37–43
Statistical process control (SPC), 20, 139, 148,
        195
Statistical tools, 8, 10, 17, 101, 172

order of usage, 20
    software, 28, 146–147
    training, 139–142, 145, 146–147
    21-step integration of, 18–21, 184–188,
        191, 197–200, 213–219
    See also Metrics; *specific tools*
Steger, Joseph A., 228
Sternberg, Robert J., 226, 230, 231, 234, 235
Stock price, as performance measure, 95, 96
Stookey, S.D., 225
Strategic business unit (SBU), 126
Strategic goals, 121–122, 136
Suppliers, 22, 97, 98
Surveys
    customer, 118–119, 120
    employee, 87–90
Svenson, Raynold A., 96

Taguchi, Genichi, 13
Takeuchi, Hirotaka, 233
TARGET (Truth, Accountable, Respect,
        Growth, Empowered, Trust), 167
Tatsuno, Sheridan M., 228
Taylor, D.W., 231
Team members, 125–126, 127
Teams
    communications, 134
    cross-functional, 170
    facilitators, 128–129, 160, 167
    motivation, 166–171, 238
    relationship patterns in, 167
    roles and responsibilities, 111, 124–129,
        136, 164–166
    size of, 7
    synergistic, 132
    See also Black Belts; Employees
Technology, and innovation, 228
Theory of Constraints, 142
30,000-foot level control chart, 20–21, 59–65,
        66, 79, 80, 156, 175, 185, 198, 215,
        219
Thor, Carl, 93, 94, 96
3M, 131
Time series data, 85, 185, 198, 215
Titanium Metals Corporation (TIMET),
        project charter of, 163–166

Tomasko, Robert M., 96
Tools. *See* Metrics; Statistical tools; *specific tool*
Total Quality Management (TQM), 5
  Conway's approach to, 11, 15
  Crosby's approach to, 12–13, 15
  Deming's approach to, 8–11, 15
  elements for success, 94
  impact on profitability, 15
  Juran's approach to, 11–12, 15
  project selection, 142
Training, 7, 60, 111, 113, 139–148
  computer software, 28, 140, 146–147
  connection to compensation, 132
  coursework, 28, 144–146
  at Eastman Kodak, 34–35
  factors in success, 34–35
  first wave of, 151–152
  goal of, 140
  in individual tools, 106, 139–140
  instructor qualifications, 28–29
  internal *vs.* external, 147–148
  mandatory, 117
  at Motorola, 32
  project-based, 107, 149
  and project selection, 142–143, 151
  return on investment, 26–27, 132
  by Six Sigma provider, 108
  strategies, 140–144
Transactional applications. *See* Service/transactional applications
Travel distance, 23
21-step integration of tools, 18–21
  development applications, 213–219
  manufacturing applications, 184–188, 191
  service/transactional apllications, 197–200

Upper specification limit (USL), 72

Variability
  between-and within-part, 191–192
  common cause, 58–59, 60, 61, 79, 82, 159, 188, 191, 195
  sources of output variability, 199, 208, 217
  special cause, 58–59, 60, 62, 82, 83
Variance components analysis, 19, 20, 64, 139, 186, 187, 199, 217
Variance-reduction strategies, 13
Variance report, 47

Wait time, 23
Walton, Mary, 10
Waste elimination, 7, 11, 23, 47
Waves, Six Sigma deployment in, 112
Weaver, Richard, 128, 130, 133, 167
Website traffic, increasing, 205–208
Weibull distribution, 90–91
Welch, Jack, 6, 32–33, 36, 97, 117, 121, 132, 135
Western Electric, Inspection Statistical Department of, 11–12
Wheeler, Donald J., 191
Wiener, Norbert, 225, 228, 233
Wiggenhorn, Bill, 31–32
Wooldridge, Adrian, 8
Wortman, William L., 46, 48, 49
Wyman and Gordon, 158, 166, 170

Yield
  final, 76
  rolled throughput (RTY), 18, 76–78, 79, 143, 185, 190, 191, 197, 214

Zero defects concept, 12
Zinkgraf, Stephen A., 17